Critically Engaging Participatory Action Research

This timely and informative book reasserts the value of Critical Participatory Action Research (CPAR): an approach to participatory action research (PAR) that is informed by critical theories attending to questions of privilege and power, and that generates collaborations focused on challenging structural inequality.

The authors, writing explicitly from Minority World perspectives, are experienced researcher-practitioners who have worked with communities in the UK, USA, South Africa, Australia, India, and Colombia over many years. They offer an assessment, exploration, and illustration of CPAR at this point in time, outlining how the approach has evolved over time and space. Exploring its roots in strands of critical thought, including postcolonialism, anti-imperialism, feminism, antiracism, queer theory, and Indigenous ontologies, the book asks how PAR is being critically re-engaged to maintain its commitment to greater justice and transformational change. Each chapter provides a rich case study of how these theories inform current collaborations and offers reflection on the entanglements of power that come with attempting CPAR in different institutional and geopolitical contexts. Their examples show that critical interrogation of PAR practices may lead to innovative and impactful outcomes for those involved, as well as new theoretical and substantive research findings.

The collection will be of especial interest to students and researchers across the social sciences and humanities, as well as those working outside universities, who are interested in developing or extending their use of CPAR.

Sara Kindon has worked for thirty years in community-based, participatory projects with Indigenous communities, women, young people, migrants, and former refugees. She has worked in a range of places, including Costa Rica, Indonesia, Aotearoa, and Oceania. Since 2006, she has provided research support to refugee-background communities and refugee-led organisations advocating for educational equity in the tertiary sector, improved service delivery, and more holistic approaches to refugee resettlement in New Zealand. Using creative and arts-based approaches, this work informed the establishment of the New Zealand National Tertiary Network to Support Refugee Background Learners and the New Zealand government's new Community Organisation Refugee Sponsorship Programme. She is the first female Professor of Human Geography and Development Studies at Te Herenga Waka Victoria University of Wellington, as well as a wife, mother, occasional dancer, and poet.

Rachel Pain is a Professor of Human Geography at Newcastle University in the UK. Her research focuses on violence, fear, and trauma, with a particular interest in gender-based violence, and analyses connections between intimate, community, and international scales. Her work is informed by feminist theory and participatory action research. She collaborates on this research with public and voluntary sector organisations and survivor groups.

Mike Kesby is a Senior Lecturer in Geography at the University of St Andrews, Scotland, UK. His research focuses on a range of health issues in Southern and Eastern Africa (HIV, sexual health, gender relations, and antimicrobial resistance in UTIs). His work is informed by participatory action research, feminist, poststructuralist, and new materialist theory. He uses participatory video and other creative methods in collaborations with a range of academic, governmental and non-governmental organisations, and grassroots citizens.

Critically Engaging Participatory Action Research

Edited by
Sara Kindon, Rachel Pain, and Mike Kesby

LONDON AND NEW YORK

Designed cover image: Getty Images / 3d Artwork Wallpaper

First published 2025
by Routledge
4 Park Square, Milton Park, Abingdon, Oxon OX14 4RN

and by Routledge
605 Third Avenue, New York, NY 10158

Routledge is an imprint of the Taylor & Francis Group, an informa business

British Library Cataloguing-in-Publication Data
A catalogue record for this book is available from the British Library

Library of Congress Cataloging-in-Publication Data
Names: Kindon, Sara Louise, editor. | Pain, Rachel, editor. | Kesby, Mike, 1966- editor.
Title: Critically engaging participatory action research / edited by Sara Kindon, Rachel Pain and Mike Kesby.
Description: Abingdon, Oxon ; New York, NY : Routledge, 2024. | Includes bibliographical references and index. |
Identifiers: LCCN 2024018211 (print) | LCCN 2024018212 (ebook) | ISBN 9780367023041 (hardback) | ISBN 9780367023058 (paperback) | ISBN 9780429400346 (ebook)
Subjects: LCSH: Action research. | Social sciences–Research. | Participant observation.
Classification: LCC HM571 .C75 2024 (print) | LCC HM571 (ebook) | DDC 001.4–dc23/eng/20240513
LC record available at https://lccn.loc.gov/2024018211
LC ebook record available at https://lccn.loc.gov/2024018212

ISBN: 978-0-367-02304-1 (hbk)
ISBN: 978-0-367-02305-8 (pbk)
ISBN: 978-0-429-40034-6 (ebk)

DOI: 10.4324/9780429400346

Typeset in Sabon
by Taylor & Francis Books

Contents

Illustrations

Figures

Tables

Contributors

Kye Askins has worked on a range of social and environment projects over the last 30 years, initially as a practitioner with homeless and mental health charities, then in the community composting field. These experiences were the foundation for taking a PAR approach in PhD and academic research, which focussed on issues of race, gender, class, and environmental justice with asylum seekers and local communities in England and Scotland. She left academia in 2019 to pursue low impact living beyond the wage economy, volunteering with organic, zero waste, and social justice initiatives as an everyday activist.

The Bawaka Collective is an Indigenous and non-Indigenous, human-more-than-human research collective. It includes Bawaka Country, Laklak Burarrwanga, Ritjilili Ganambarr, Merrkiyawuy Ganambarr-Stubbs, Banbapuy Ganambarr, Djawundil Maymuru, Kate Lloyd, Sandie Suchet-Pearson, and Sarah Wright. Bawaka Country is the diverse land, water, human, and nonhuman animals, plants, rocks, thoughts, and songs that make up the Yolŋu homeland of Bawaka in North East Arnhem Land, Australia. Laklak, Ritjilili, Merrkiyawuy, and Banbapuy are four Indigenous sisters, elders, and caretakers for Bawaka Country, together with their daughter, Djawundil. Sarah, Sandie, and Kate are three non-Indigenous human geographers from the University of Newcastle and Macquarie University who have been placed into the family as granddaughter, sister, and daughter. The collective has worked together since 2006.

Sowmyaa Bharadwaj, over the last fifteen years, has been involved with several participatory research studies, monitoring designs, assessments, and evaluations across a range of thematic areas of development, including livelihoods, sanitation and hygiene, gender and sexuality, climate change, urban development, and rights. Her work with urban and rural communities has focussed on the use of participatory methods to unpack issues of equity and governance in different spheres. She is an experienced facilitator, a practitioner of participatory approaches, and has been involved in various trainings and capacity-building initiatives with a range of government, nongovernment organisations, and community groups across India.

Laklak Burarrwanga is a Datiwuy Elder, Caretaker for Gumatj, and eldest sister. As such she has both the right and the cultural obligation to share certain aspects of her knowledge and experiences with others. She has many decades of experience sharing this knowledge with children, through years of teaching in the community and at Bawaka, and through writing and translating books for the Yirrkala Community School. She has also communicated this knowledge through weaving, painting, and printmaking, and as a member of the National Museum of Women in Art. She established her family-owned tourism business, Bawaka Cultural Experiences (BCE), and through this business shares her knowledge with tourists, including government staff in cross-cultural programs. Laklak received an Honorary Doctorate from Macquarie University in 2016 in recognition of her profound knowledge and passionate commitment to cross-cultural communication and sharing of knowledge. Dr Burarrwanga continues to lead the Bawaka Collective in different ways since her passing.

Caitlin Cahill is a community-based urban and youth studies scholar, environmental psychologist, and organiser from New York City, USA. She is grateful to collaborate with a community of changemakers to create a more liveable, just, caring, and free world for all of us. Caitlin's participatory praxis and research focus on the everyday, intimate experience and struggle against racial capitalism as it concerns gentrification, immigration, education, and state violence. Projects include Re:Play, the Bushwick Action Research Collective, Growing Up Policed, and Makes Me Mad! (with the Fed Up Honeys). In Utah, Caitlin cofounded the Mestizo Arts & Activism Collective. Caitlin is an Associate Professor of Urban Geography & Politics and co-coordinator of the Social Justice/Social Practice Minor at Pratt Institute.

Anusha Chandrasekharan is a journalist by training, with ten years' experience in the development sector. She has worked on several research and capacity-building assignments that aim to enable community participation in different stages of a development project in contexts including gender and sexuality, disability, child rights, and modern forms of slavery. She has used participatory methods with communities marginalised by sex, gender, occupation, caste, class, religion, and ability and has used participatory video and digital storytelling as tools of advocacy, in inclusive urban planning, with survivors of trafficking, and sexual minorities.

Dheeraj has a decade of experience in community institution building; community resource strengthening; developing and strengthening sustainable implementation systems on livelihoods enhancement; creating livelihood prototypes; access to rights and entitlements; gender equality; and access to basic services in rural and tribal belts. He has engaged in capacity building of communities as well as organisations in the realm of gender, rights and entitlements, and livelihood programmes. He had also been instrumental in providing capacity building support to government and nongovernment organisations in the realm of livelihoods enhancement, supply chain strengthening, business responsibility, and participatory methods and approaches, including grassroots engagement.

Banbapuy Ganambarr grew up at Guluruŋa. She is a bilingual student who completed her degree at Bachelor College through the Northern Territory University. Banbapuy has been a senior Indigenous teacher at Yirrkala School. Her work as an author, artist, weaver, and teacher has allowed her to influence curriculum and teaching methods, conduct professional development for teachers at the school, and to stand up and explain community needs and goals to government departments and officers. She works closely with her family, supporting them in running a cultural tourism business.

Ritjilili Ganambarr is the second eldest daughter, a Datiwuy elder, and caretaker for Gumatj. She has worked hard on health issues in the community and is passionate about working with mothers and children, teaching and educating them that strong mothers create strong children. She is a weaver and writer/illustrator. She is a coauthor of two books, *Welcome to My Country* and *Weaving Lives Together*, as well as a book chapter for teachers on sustainability. She is an illustrator of *Nganapu Nguli Marrtji Diltjiyi (We Go Out to the Bush)*, a dual-language book written in Yolŋu Matha and English. Ritjilili works with her family's highly successful Yolŋu owned-and-run Indigenous tourism business (Bawaka Cultural Experiences). She's regularly presented workshops to non-Indigenous participants on Yolŋu culture and land. She continues to care for the Bawaka Collective and play an important leadership role since her passing.

Merrkiyawuy Ganambarr-Stubbs is a proud Yolŋu woman and leader from North East Arnhem Land. She has written six books. Her children's books are written in Yolŋu Matha for use in primary schools as Walking Talking texts. She plays an important role in the bilingual education movement in Arnhem Land, working with Yolŋu Elders to develop both-ways learning. She has developed a series of Yolŋu curriculum materials currently in use in Arnhem Land. She has also been an important voice for Yolŋu rights. In this capacity, she has regularly appeared on television and the internet. She is currently Director of Yolŋu Education and Co-Principal at Yirrkala School. Her vision is that every child is appreciated, and that every child knows that dreams are possible.

Mike Kesby is a Senior Lecturer in Geography at the University of St Andrews Scotland UK. His research focuses on a range of health issues in Southern and Eastern Africa (HIV, sexual health and gender relations and most recently antimicrobial resistance in UTIs). His work is informed by participatory action research and feminist, post-structuralist, and new materialist theory. He uses participatory video and other creative methods in collaborations with a range of academic, governmental and non-governmental organisations, and grassroots citizens.

Sara Kindon has worked for thirty years in community-based, participatory projects with Indigenous communities, women, young people, migrants, and former refugees. She has worked in a range of places, including Costa Rica, Indonesia, Aotearoa, and Oceania. Since 2006, she has provided research support to refugee-background communities and refugee-led organisations advocating for educational equity in the tertiary sector, improved service delivery, and more holistic approaches to refugee resettlement in New Zealand. Using creative and arts-based approaches, this work informed the establishment of the New Zealand National Tertiary Network to Support Refugee Background Learners and the New Zealand government's new Community Organisation Refugee Sponsorship Programme. She is the first female Professor of Human Geography and Development Studies at Te Herenga Waka Victoria University of Wellington, as well as a wife, mother, occasional dancer, and poet.

Kate Lloyd is a teacher, researcher, and Professor at Macquarie University, Sydney, Australia. Kate's work focuses on several projects that take an applied, action-oriented, and collaborative approach to research characterised by community partnerships, co-creation of knowledge, and an ethics of reciprocity. She has experience doing research and working with government and community organizations in Vietnam, Lao PDR, PNG, and northern Australia. Over the past 17 years, she has worked with the Bawaka Collective to make contributions around Indigenous knowledges and collaborative methodologies.

John Marnell is based at the African Centre for Migration and Society, University of the Witwatersrand, South Africa. His most recent book publication is *Seeking Sanctuary: Stories of Sexuality, Faith and Migration* (2021). He is also the co-editor of *Queer and Trans African Mobilities: Migration, Asylum and Diaspora* (2021), which won the 2023 ASR Best Africa-focused Anthology or Edited Collection Prize. John is the co-convenor of the African LGBTQI+ Migration Research Network (ALMN).

Djawundil Maymuru is a Maŋgalili woman, raised by a Gumatj elder. She is a Yolŋu mother, grandmother, and great-grandmother from the beautiful homeland of Bawaka in North East Arnhem Land. She is a co-author of three books, *Welcome to My Country, Weaving Lives Together* and *Songspirals*. Djawundil works with Bawaka Cultural Experiences, a highly successful Yolŋu-owned-and-run Indigenous tourism business. As a key member of the business, she works with visitors to Bawaka to share life at Bawaka with them, helping them understand and respect Yolŋu culture and land. She is a college graduate, a bilingual student, and has been on the board of Laynhapuy Homeland Association and Lirrwi Tourism Corporation. Djawundil has also been invited to share Yolŋu knowledge at conferences and seminars in New Zealand, Canberra, Sydney, Boston, Bellingen, and Newcastle, Australia.

Rachel Pain is a Professor of Human Geography at Newcastle University in the UK. Her research focuses on violence, fear, and trauma, with a particular interest in gender-based violence, and analyses connections between intimate, community, and international scales. Her work is informed by feminist theory and participatory action research. She collaborates on this research with public and voluntary sector organisations and survivor groups.

Amy Ritterbusch has led social justice-oriented participatory action research initiatives with street-connected communities in Colombia for the last decade, and recently in Uganda. Her work involves the documentation of human rights violations and forms of violence exerted against homeless individuals, sex workers, drug users, and street-connected children and youth, and she works with the subsequent community-driven mobilizations to catalyze social justice outcomes within these communities. Throughout her research and teaching career, she has explored different approaches to engaging students and community leaders through critical and responsible interaction between classroom and street spaces in Colombia and Uganda, using the lens of social justice-oriented PAR. Her research has been funded by the Open Society Foundations, the National Science Foundation, the Fulbright US Program, and other networks promoting global social justice. She is currently Associate Professor of Social Welfare, UCLA Luskin School of Public Affairs in California, USA.

Jackie Shaw is Research Fellow at the Institute of Development Studies (IDS), United Kingdom, who uses visual and performative methodologies to drive participatory action research and community-led change processes. She is a Participatory Video pioneer, who co-authored the definitive guide, *Participatory Video: A Practical Approach to Using Video Creatively in Group Development Work* (1997), and has extensive experience collaborating with marginalised groups in diverse community, development, and health contexts. Her social psychology PhD built nuanced understanding of empowerment practices. During the Participate programme, she collaborated with partners in India, the Palestinian West Bank, Kenya, and Indonesia to bring the reality of poverty to United Nations (UN) decision makers. Her recent research has explored green-energy conflict at the rural margins in Kenya, and how to build accountable relationships between participants and influential duty-bearers in highly inequitable contexts in India, Egypt, Ghana, Uganda, and South Africa. She is currently researching disability inclusion with consortium partners in six countries, and collective healing processes in Africa.

Brett Stoudt, PhD is an Associate Professor in the Psychology Doctoral Program at the City University of New York, Graduate Center where he is the former head of the PhD program in Critical Social/Personality and Environmental Psychology, and currently serves as Psychology's Deputy Executive Officer of the Graduate Center Campus. He is also a faculty member in the Social Welfare and Urban Education doctoral programs. Dr. Stoudt has worked on numerous participatory action research projects with community groups, lawyers, and policymakers, nationally and internationally. He is currently the Associate Director of the Public Science Project and is also actively involved with Communities United for Police Reform as a steering committee member.

Sandie Suchet-Pearson is Professor in Human Geography at Macquarie University in Sydney, Australia and currently holds a Future Fellowship. Her research and teaching experiences over the last 25 years have been in Indigenous rights and environmental management. She's worked on Cape York Peninsula on community development in the context of a major mining operation, examined the strategies used by Indigenous peoples and local communities to assert their rights in wildlife management in Canada and southern Africa, and her current work focuses on Indigenous self-determination in the context of cultural tourism in North East Arnhem Land, northern Australia and caring-as-Country in western Sydney.

Nina Woodrow is a researcher, educator and creative practitioner who uses arts-based methods to explore cultural dimensions of contemporary sociality. Her special skill is in the design of community-engaged methodologies to respond to the ethical challenges of urban diversity. She draws on a history of community arts and cultural development practice, and a multidisciplinary academic foundation in performance theory, sociology, cultural geography, and education. Nina holds a PhD from Queensland University of Technology, and she currently teaches arts and humanities in community and university settings in and around Brisbane, Australia.

Sarah Wright is a human geographer and Professor from the University of Newcastle, Australia. She holds a Future Fellowship that aims to attend to Weather Cultures, including their expression through songs, songlines, and stories to understand the way weather actively and affectively co-constitutes people and place. She has worked with community groups and NGOs for over 30 years in Australia, Mexico, Kenya, South Africa, Cuba, and the Philippines. She works in the Philippines with a network of subsistence organic farmers and is part of Yandaarra, a Gumbaynggirr-non-Gumbaynggirr in Australia. For the past 17 years, she has worked closely with inspirational Yolŋu women from North East Arnhem Land and has been placed into the family at Bawaka homeland.

Acknowledgements

A book such as this involves many people whose names appear and do not appear on its pages.

We are grateful to all the contributors here who have borne with us through the gestation of this volume. Your insightful and principled work inspires us, and your collegiality and understanding have been wonderful.

We also thank those contributors who were unable to join us due to other demands on their time and energies because of the pandemic. We appreciate all you do in advancing social justice and participatory work within your institutions and their wider communities.

Our many postgraduate students and colleagues have inevitably contributed to our own thoughts and work. We hope this volume supports your ongoing scholarship and activism.

In helping us to bring everything together, we also want to acknowledge the work of Dr Maja Zonjić (images) and Anna Rogers (copyediting). Your attention to detail and consistency, plus good humour, has been much appreciated.

And finally, at Routledge, Andrew Mould, Egle Zigaite, Claire Maloney, and Charlene Price have been great to work with. We thank you for your ongoing patience and encouragement.

1 Critically Engaging Participatory Action Research

Sara Kindon, Rachel Pain, and Mike Kesby

Introduction

The explosion of interest in Participatory Action Research (PAR), which prompted our previous book (Kindon, Pain, & Kesby, 2007), has continued. We estimate that PAR constitutes the fastest growing research approach within academic institutions across the social sciences, which is the context from which we write. Across the world, it is an even more popular approach for local communities to generate shared knowledge and pursue change. Taking many different forms, the approach is now mature and widely used.

PAR is an approach to social enquiry that emerges from many different peoples' long-term and embedded struggles around the world. It presents a challenge to both positivist and postpositivist inquiry (Heron & Reason, 1997; Riley & Reason, 2015), because a participatory onto-epistemology "sees neither a world of separate things (a la positivism) nor a world constructed predominantly by the human mind or systems of social reproduction (a la various relativist perspectives)" (Gayá, 2021, p. 171). It understands that "knowers, and that which is known, and unknown, are entangled in relationship with one another, in localized and material ways, informed by diverse, multiple ways of knowing" (Gayá, 2021, p. 171). Of course, this is also true of social constructionist perspectives. What distinguishes PAR is that, ontologically, people other than academics are recognised as situationally reflective analysts of the world, who have particular expertise in relation to researching their own lives and, using this knowledge, they are able to take action to change their lives for the better.

Epistemologically, these more fully *social* approaches to the production of knowledge (mrs c kinpaisby-hill, 2011), driven by those who have been traditionally positioned as subjects of research, are more likely to produce knowledge around which validity, utility, and truth claims are more robust. The call for researchers to say and do 'nothing about us without us', adopted from disability activism, has scientific as well as political weight in the search for collaborative approaches to imagining alternatives to present-day injustices, challenges, and unwanted conditions (Gergen, 2015). The strength of PAR lies in its ability "to facilitate the intersections of theory, practice and politics between participants and researchers in a diversity of contexts" (Kindon et al., 2007, p. 3). As Patricia Carolina Gayá (2021) notes, this approach enables "multiple ways of knowing [to] connect and build on one another in the context of collaborative action and participatory research involving individuals, groups and communities outside of conventional research roles or academic contexts" (p. 172).

For many, this mode of research offers potential for communities to generate knowledge that is more representative of their own needs and perspectives, and to develop

DOI: 10.4324/9780429400346-1

forms of knowledge production that are more democratic and action-oriented, and which exceed neoliberal forms of academic labour. It also offers a practical means by which to negotiate the reality that all knowledge is partial and situated, and that all epistemologies are necessarily incomplete. As Gayá (2021) suggests, PAR involves a will to be epistemologically expansive and "to 'grant equality of opportunity' to different kinds of knowledges and epistemological positions, with a view to maximising their respective contributions" (citing de Sousa Santos, 2014, p. 182) in the quest to investigate and shape the world.

As previously dominant epistemologies and knowledge claims have been shown to be partial and open to contestation, academics and their institutions have increasingly been called to account for the privilege that has accrued from colonialism, and for their entanglement with the harm wrought by elitist and extractive research processes on disenfranchised and marginalised people (Fine & Torre, 2019; Lenette, 2022; McKittrick, 2021; People's Knowledge Editorial Collective, 2016; Tuck & Yang, 2012; Tuhiwai Smith, 1999). One consequence – a double-edged one – for academic research is that greater value is now accorded to public scholarship that facilitates public participation in knowledge production.

The promotion and uptake of participatory approaches has also occurred at the same time as a set of contradictory shifts in academic institutional practices associated with the privatisation and marketisation of higher education, increasing competition for research funding and growing worker precarity (Burton & Bowman, 2022; Morley, 2015; O'Keefe & Courtois, 2019; Parker, Martin-Sardesai, & Guthrie, 2023). There has also been a slow and very partial shift in how some institutions expect researchers to practice ethics. This includes the need for ethics to be negotiated throughout the research process (Kindon & Latham, 2002), though it is still common for ethics review boards to define what is meant by ethics (Tuck & Guishard, 2013), and to present difficulties for long-term and collaborative projects driven by community concerns (see Cahill, Chapter 2).

In parts of the Minority World, for example the United Kingdom, the rise of audit cultures designed to monitor academic performance, which stimulate research productivity and account for public investment in research, has called for the demonstration of research impact (Pain et al., 2011). This requirement has had significant consequences for how participatory practice is understood and used. To some, PAR might seem more possible than ever since its goals ostensibly parallel some of those now authorised by institutional vision statements and metrics. However, the more that PAR has been branded, systematised, and packaged as a set of 'tools' or 'methods' that can deliver innovation, involvement and impact, the more the politics of PAR (i.e., its nomenclature, radical history, and deeper structural change work) has been decentred and depoliticised within notions of 'co-production' or 'co-design' (see Kindon, 2010). PAR approaches started out as radical, organic, and context-specific responses to the power imbalances and unjust outcomes of much conventional research. Yet generic terms such as *co-design* neither assume nor require any particular onto-epistemological position, instead reflecting a politics of knowledge production that has been captured by the economics of institutional reproduction.

As a consequence, PAR approaches have often in past decades come to resemble technocratic 'fixes' to address short-term problems that can be marketed at the end of a project funding cycle. Therefore, from our perspective, too much of what now purports to be PAR – participatory research, co-produced or co-designed research, within academic settings – lacks criticality. Simply arguing for a return to some notion of a purer, more organic PAR offers no panacea to this crisis. PAR, as we have noted elsewhere, is always, already, enmeshed with power as a situated, spatial work-in-progress (Kesby, Kindon, & Pain, 2007). There is no gold standard of PAR praxis (Kesby, Kindon, & Pain, 2005), as its use and its ability to effect meaningful change for people, places or policies are highly variable, contested, and unpredictable (Pain, Kindon, & Kesby, 2007). Nonetheless, and as is signalled by this book, this polyvalent research approach can and *should* be underpinned by the political commitments that are manifested in certain strands of critical theory and practice, as we discuss below.

In a recent volume (Burns, Howard, & Ospina, 2021), PAR was likened to "a dynamic landscape, a watershed with its many hydrographic basins whose rivers and their tributaries flow into the same sea" (Ospina, Burns, & Howard, 2021, p. 8). As geographers, we love geomorphological and landscape metaphors, but when it comes to PAR, these metaphors are too bounded and linear, too physical and predictable, to be helpful. The flow of PAR is socially and politically determined, not a function of gravitational laws (indeed, movement has often been 'uphill'). Streams of thought and action span continents (with the historic and contemporary flows from Majority to Minority World too often neglected, and all channels do not necessarily lead in the same direction or to the same inevitable endpoint. PAR is a fragile, constantly evolving relational project that needs work, renewal, and praxis to maintain its effects and affects. Its tensions and contradictions require continual nurturing and attention to facilitate ethical and just practices.

For us, PAR is less a braided river and more a series of ongoing, unfinished improvisations. Such improvisations may draw on inherited 'scripts', 'genres', and 'techniques', but when most effective, also actively encourage situated, relational, and embedded innovation. They refuse a neat division between 'directors', 'players', and 'audiences', or between 'backstage', 'stage', and 'front of stage'. They seek to both represent and change the worlds we inhabit, as Augusto Boal's work shows (1979, 2006). Crucially, these improvisations – and the doing-living-feeling of PAR – can (even *should*) involve antagonistic discussion and unresolved contradictions as much as it can enable mutual agreement and resolution.

PAR is therefore inherently emotional. Shared senses of outrage, anger, bewilderment, fear, grief, or urgency are often the catalysts for collaborative inquiry and action (Cahill, 2004). Such emotions often provoke questions that generate new insights, fuel, and sustain long-term commitments, and effect change in other people or within systems. In fact, it is common to experience improvising PAR as uncomfortable, challenging, difficult, imperfect, or unfinished when there is a deep commitment to shared responsibility and the genuine negotiation of power. So much comes down to how PAR is deployed, and how power and privilege are acknowledged and mobilised.

In this regard, focusing on *Critical* PAR (CPAR) is both timely and essential. As we go on to describe in this introductory chapter, Critical PAR is an umbrella term for PAR informed by several theoretical approaches that all attend fundamentally to questions of privilege and power. In so doing, CPAR:

- denaturalises and reframes oppression and injustice;
- challenges epistemological standpoints of empirical scholarship that are positioned/disguised as objectivity;
- acknowledges privilege as a set of intersecting material and sociopsychological conditions, which can be both enabling and constraining; and
- moves beyond passive empathy to collective responsibility, solidarity, and action.

By framing this collection within and about CPAR, we offer our assessment of where PAR and CPAR are now. The chapters that follow share insightful, analytical reflections on attempts to practice a PAR that is both critical and critically engaging. The contributors attend to various related issues around theory, arts-based praxis, and the limits of participation, and take care to review the foundational frames and dilemmas of praxis. Later in this introductory chapter, we draw together some key themes from across the book that underpin the *critical* dimensions of PAR moving forward.

We therefore want this book to contribute to the calls of other commentators to 'press pause' in the uncritical proliferation of all things participatory in research. Our purpose is not to introduce, explain or instruct how to do CPAR, especially given our critiques of the technocratisation and methodological fetishism that have beset PAR noted above. (In addition, numerous practical guides already exist.) Instead, we share reflexive accounts about the politics of CPAR, as a resource for others and to support more critical future uses of PAR.

To realize these aims, this book answers three main questions:

1 How has PAR evolved over time and space, and where does CPAR fit within this genealogy? This question is engaged with in this chapter.
2 In what ways has or could PAR be critically engaging? Our contributors provide some rich examples in the chapters that follow.
3 How can we critically (re)engage PAR to maintain its orientation to greater justice and transformational change? We address this in our discussion of the book's cross-cutting themes later in this chapter, and the question is also addressed in the chapters that follow.

In the rest of this chapter, we review CPAR's historical evolution and current diversity. Within our review, we offer some comment on how CPAR is both similar and significantly different to many instances of research that are oriented towards co-production or co-design, or otherwise claim to be participatory. We then position the book, including acknowledging what is missing from it, before identifying the book's key themes, drawing on insights from the chapters that follow and the connections between them.

The Evolution of Critical Participatory Action Research

> Do not monopolise your knowledge nor impose arrogantly your techniques, but respect and combine your skills with the knowledge of the researched or grassroots communities, taking them as full partners and co-researchers. Do not trust elitist versions of history and science which respond to dominant interests, but be receptive to counter-narratives and try to recapture them. Do not depend solely on your culture to interpret facts, but recover local values, traits, beliefs, and arts for action by and with the research organisations. Do not impose your own ponderous scientific style for communicating results, but diffuse and share what you have learned together with the people, in a manner that is wholly understandable and even literary and pleasant, for science should not be necessarily a mystery nor a monopoly of experts and intellectuals.
>
> (Fals-Borda, 1995, np)

As we noted above, PAR approaches have diverse but overlapping points of origin. Orlando Fals-Borda (2006a) pointed to several strands or schools emerging out of particular intellectual and activist traditions and contexts (see Kindon et al., 2007 for an earlier summary). While these points of origin share common features, they may look quite different in terms of their politics and practices on the ground.

Foundations

What is now recognised as the *critical* tradition of PAR developed in several countries in the Majority World during the 1960s and 1970s, particularly in Africa, Latin America, and the Indian subcontinent (Glassman & Erdem, 2014). This work represented a new decentralised epistemology of practice that was grounded in people's struggles and local knowledges. Its development was closely tied to anticolonial independence struggles and to social movements for civil rights and peace. As practices evolved through the 1980s and 1990s, and with the rise of the women's movement, calls for Indigenous self-determination, and environmental concerns, the assertion of feminist and Indigenous knowledges played increasingly important roles.

While this history is rich and diverse, some key thinker-practitioners stand out as influencing PAR's evolution in particularly important ways associated with their approach, nomenclature, language, and politics (Table 1.1). Our choices here are inevitably partial. We foreground them not to solidify PAR's history, but to restate some important strands that underpin present day deployments of participatory research and reassert their legacies at a time of overly sanitised accounts of co-production.

PAR, then, originated primarily in Majority World contexts, and has been shaped by Indigenous radical politics, theory, and practice oriented toward social and systemic change. As signposted in Table 1.1, these thinker-practitioners have drawn on several strands of critical thought, including postcolonialism, anti-imperialism, feminism, anti-racism, and Indigenous ontologies. This early work continues to inform ground-up research by and with communities, on issues that communities define, which may or may not be identified or labelled as PAR (Coombes, Johnson and Howitt, 2014; Coombes, 2017; Bawaka Country, Chapter 3).

While this body of work is diverse, and not homogenous within Majority or Minority Worlds (see Moyo, 2020), together it lays the foundation stones for Critical PAR. Today CPAR spans all parts of the world, arising from radical marginalised perspectives, particularly feminist, Indigenous, and Black praxis (Kovach, 2015; Lenette, 2022). CPAR continues to challenge more liberal deployments of PAR and other forms of participatory practice, as we now go on to discuss.

The Move to the Minority World

Only because of these early developments did PAR evolve in the more affluent countries of the Minority World during the 1980s and 1990s, particularly through its intertwining with strands of Action Research and later Community Based Participatory Research originating in the United States of America, and the integration of participatory approaches into international aid and development work.

Narrations of PAR from within the Minority World often describe it as stemming from Karl Lewin's 1946 development of Action Research (AR) to inform social and organisational change. However, this is properly viewed as a separate development, with variants of PAR having a much firmer basis in philosophies of Majority World origin (Glassman & Erdem, 2014), as we have described above. That said, Lewin's AR did set out some key practices that have been adopted in Minority World PAR. For example, his description of an "iterative process of interplay between researcher and participants in which activities shift between action and reflection" (Fisher & Ball 2003, pp. 209–210) has become commonplace in certain academic fields such as social psychology and

Table 1.1 Some Key Thinker-Practitioners Who Created and/or Influenced the Development of Participatory Action Research

Mahatma Gandhi (India, 1868–1948)
Mahatma Gandhi was a leader of India's independence movement. His work in anticolonial independence struggles, and the principles and practice of his activism had wide-ranging influence on the development of PAR. In particular, Gandhi (1928) developed a method of nonviolent resistance, involving noncooperation and peaceful tactics to fight injustice. He called on the public, especially poor and oppressed people, to draw on satyagraha ("soul force" or "holding onto truth"), power from within, to persuade opponents with patience and compassion.
Paulo Freire (Brazil, 1921–1997)
Paulo Freire was an emancipatory educator who developed community-based research processes to support people's participation in knowledge production and social transformation; an early form of what later became known as PAR. He worked with illiterate students for whom education was crucial in challenging their marginalised status, and he developed a critical or radical pedagogy in his book *Pedagogy of the Oppressed* (1972). Through a process of *conscientização*/conscientization, poor and marginalised groups develop a heightened awareness of the forces affecting their lives, and then use this as a catalyst to inform political action. Crucially, he theorised knowledge not as transferred from one (expert) group to another (disempowered) group, but as coming from within. For Freire, education and knowledge production therefore involve those traditionally considered teachers/students or researchers/researched working alongside each other in more equitable knowledge exchange and theory-building: "The silenced are not just incidental to the curiosity of the researcher, but are the masters of inquiry into the underlying causes of the events in their world. In this context research becomes a means of moving them beyond silence into a quest to proclaim the world" (Freire, 2007, p. 30–31). His ideas provided great inspiration to others in the Majority World who were dissatisfied with the legacies of colonisation, modernistic development interventions, and positivistic research paradigms.
Orlando Fals-Borda (Colombia, 1925–2008)
Orlando Fals-Borda was a rural sociologist who worked with peasants and fisherpeople in coastal Colombia from the 1960s. Combining his academic and political commitments, he named the research approach he developed with local people "participatory action research," recognising that knowledge is power and that liberation must involve dismantling the duality between researcher and researched. He argued that PAR was a set of attitudes and values informing praxis and not simply a research methodology, but a "philosophy of life" that would turn its practitioners into "*sentipensantes*"/"thinking-feeling" persons (2006b) and inspire revolution. His approach, including "imputation," where he "seized hold of historical information and gave it body through his empirical imagination" (Rappaport, 2020, p. xxi), went on to have much wider influence in Colombia and beyond.
Marja-Liisa Swantz (Finland/Tanzania, 1926–)
Marja-Liisa Swantz is a Finnish scholar who lived in Tanzania from the 1950s, initially as a teacher trainer. Her work with women in the 1960s and 1970s, laid the foundations for what she called "participatory research." Her approach was in response to development research and practice of the time, which bypassed local knowledge and extracted benefits from communities without reciprocating. She was especially critical of the way women were exploited in this way, and she worked alongside local women as they integrated their own knowledge and expertise into research on the issues that mattered most to them. She argued that such research, conducted inside communities by communities, makes for less subjective findings and more beneficial outcomes (Swantz, 1985). Despite the importance of her work, the significant influence of Swantz's ideas, practices, and analysis of gender in participatory research was not initially acknowledged by the male theorists who (still) tend to receive more credit in the genealogy of PAR (Glassman & Erdem, 2014). Swantz's work therefore both set the baseline for feminist PAR and continues to provide an exemplar.

Linda Tuhiwai Smith (Aotearoa, New Zealand, 1950–)

Linda Tuhiwai Smith is a Professor of Indigenous Education. Her book *Decolonizing Methodologies* (1999) provided a searing critique of Western research practice as a weapon used in service of colonialism and neocolonialism, harming Indigenous people and their relationships with their lands and cultures: "The word 'research' is probably one of the dirtiest words in the indigenous world's vocabulary . . . it is implicated in the worst excesses of colonialism and remains a powerful remembered history" (Tuhiwai Smith, 1999, p. 1). This work, and her contributions to the development of Kaupapa Māori theory and educational research practice in Aotearoa, have been hugely influential in the development of Tribal/ Indigenous variations of, engagements with, and distinctions from PAR around the world. Typically, in these orientations, control over the research process and knowledge produced is held by Indigenous leaders who only invite collaboration from external (potentially colonising) researchers, where there is perceived strategic benefit.

Rajesh Tandon (India, 1951–)

Rajesh Tandon is recognised as the founder of community-based research (closely related to PAR) in Asia. As a development practitioner, he developed new ways of working with local people, supporting them to utilise their existing knowledge for progressive change. He went on to found the organisation Participatory Research in Asia, which supports the theory and practice of grassroots initiatives, and is now a UNESCO Chair in Community-Based Research.

Patricia Maguire (United States, 1951–)

Patricia Maguire critiqued PAR from a feminist perspective. In particular, she identified chauvinism within the development of PAR. She also called attention to previously overlooked problematic imbalances of power in communities that affected PAR's use and implementation, and sometimes resulted in the reinforcement of the (usually older masculine) status quo. Setting out an agenda for *Doing Participatory Research: A Feminist Approach* (1987), Maguire brought the ethical and political imperatives of feminist research into dialogue with PAR. Since this time, the approach has been very influential, shaping contemporary strands of Critical PAR.

educational research. However, while AR may overlap with PAR and include participatory elements (see Reason & Bradbury, 2008), research grounded in and owned by communities is generally a central tenet of PAR rather than AR. Like Lewin's AR, Whyte's (1994) version of PAR had strong roots in the social and political contexts of the United States in the mid-twentieth century, but not as a vehicle for radical change. Rather, it presented an approach focused on conflict resolution, and maintaining rather than challenging the status quo (Glassman & Erdem, 2014).

In the United Kingdom in the 1980s and 1990s, academic Robert Chambers, inspired by the work of Tandon and others in the Majority World, argued for their radical reconceptualisation of Western-led development practice. Chambers argued that poor communities be recognised as holding the most accurate and useful knowledge about their situations, and that they should be supported to generate knowledge and participate in change from which they benefit (Chambers, 1983, 1997). These principles were accompanied by methodological innovations (Chambers, 2002); essentially borrowing, developing, and adapting various hands-on visual and community development techniques from the nongovernmental organisations with whom he had been working.

Chambers branded iterations of his approach Participatory Rural Appraisal (PRA), an approach which aimed to be "a family of approaches, behaviours and methods for enabling people to do their own appraisal, analysis and planning, take their own action, and do their own monitoring and evaluation" (Chambers, 2002, p. 7). Later, Participatory Learning and Action (PLA) more closely resembled PAR as we might recognise it.

PRA and PLA included techniques such as mapping and modelling, transect walks, matrix scoring, wellbeing grouping and ranking, seasonal calendars, institutional diagramming, trend and change analysis, and analytical diagramming, all undertaken by local people. They have been used around the world in a variety of settings – including natural resources management (e.g., soil and water conservation, forestry, fisheries, wildlife, village planning, and so on), agriculture, health, nutrition, food security, and education – to encourage deliberative interaction and decision making in culturally appropriate and inclusive ways to influence more sustainable and equitable policy and practice (Chambers, 1994).

Knowledge about these methods and tools was rapidly taken up by academics aspiring to do research differently (see Kindon, 1995; Kesby, 2000; Alexander et al., 2007). Over this time, participatory approaches moved from being primarily a radical critique of mainstream positivist and extractivist research to being closer to the centre of academic practice. Participation became fashionable (Pain & Francis, 2003), in response to state policy agendas; nongovernmental organisations; changes in higher education contexts, particularly neoliberalism and marketisation; and pressure for universities to demonstrate the relevance of research and its social and economic impacts. These changing conditions both provided opportunities and deepen challenges for those committed to participatory research and to PAR (Pain et al., 2011).

Many of the tenets of feminist methodology also developed in tandem with participatory epistemologies, giving rise to more ethical, reciprocal, and relational research methods (e.g., Brydon-Miller, Maguire, & McIntyre, 2004; Cahill, Sultana, & Pain 2007; Fine & Torre, 2019; Kindon, 2003, 2012; Kindon et al., 2007; Lenette, 2022; Monk, Manning, & Denman, 2003), having some positive impacts on how at least some social science research is carried out. Meanwhile, as Banks, Hart, Pahl and Ward (2018) chart, so-called co-produced research has mushroomed within academia over the past twenty years, an overarching umbrella within which more specialized types sit: everything from collaborative ethnography to community-university action research. Banks, Hart, Pahl and Ward (2018), see co-produced research as research done together by people from different backgrounds and experiences, based on a participatory worldview, with a commitment to principles of equality and democracy, drawing on many traditions.

Nonetheless, the theory and practice of what is called *participatory* are often two quite different things, and co-production does not always attend to politics, power, and privilege, or work towards deep structural change and transformation of enduring systemic injustices. While there are many examples of profound relationships, community mobilisation, and change for good, the academic landscape is littered with other examples of "engagement", "collaborative research", "co-design", and "co-production", often sitting within the double-edged category of "research for impact" (Pain, Kesby, & Askins, 2011). Growing efforts to attend to diverse knowledges, engage deeply with people outside of academia, and foster transformation and justice, while laudable, have had mixed results for those communities (Kindon et al., 2007).

Critiques of Participation and Participatory Research

Perhaps the fiercest critical scrutiny of participatory theory and practice has been levelled at international participatory development. Here, the deployment of participatory approaches for community development in the Majority World has frequently been led by researchers and international development practitioners in or from the Minority World.

For example, Chambers's early work in this arena was criticised for continuing to reinforce forms of tyrannical control (Cleaver, 1999; Kapoor, 2002; Kothari, 2001).

In what Leal (2007) described as "the ascendancy of a buzzword," with the once-radical ideal of participation becoming "something that could serve the neo-liberal world order" (p. 539), participation over the last thirty years has become scaled up, institutionalised, and deradicalised (Cooke & Kothari, 2001; Hickey & Mohan, 2004). Widespread co-optation, in the forms of participatory development planning, budgeting, implementation, monitoring and evaluation, where community input was facilitated but power and control remained with external partners, led Cooke and Kothari to question whether reform of participatory development was even possible. In their 2001 book, they stated:

> Tyranny is both a real and potential consequence of participatory development, counter-intuitive and contrary to its rhetoric of empowerment though this may be. . . Authentic reflexivity requires a level of open-mindedness that accepts that participatory development may inevitably be tyrannical, and a preparedness to abandon it if this is the case.
>
> (Cooke & Kothari, 2001, p. 15)

Such postcolonial critiques gathered momentum, suggesting that the advocates of participation were reauthorising themselves as experts through participatory processes, reinscribing existing power relations rather than truly handing over authority to participants (Kothari, 2005; Mohan, 1999). Such professionals were critiqued for presenting as benign arbiters of neutral or benevolent processes, but simultaneously legitimising neoliberal programmes and institutions (such as development agencies) that deployed participatory approaches and/or techniques (Kindon et al., 2007; People's Knowledge Editorial Collective, 2016). Indeed, over time practitioners observe that participatory approaches often work best for communities on small projects, with critical awareness about the limits of the emancipatory claims of the methods involved (Kanyamuna & Zulu, 2022).

In perhaps the most scathing account, Kapoor (2005) accused the practitioners who took up Chamber's approaches as indulging in "narcissistic samaritanism" (p. 1206), taking on the role of benevolent, self-effacing, apparently neutral facilitators with the stated aim of winning improvements for poor communities. He highlighted how the very act of "pretending to step down from power and privilege, even as one exercises them as master of ceremony" (Kapoor, 2005, p. 1207) served to reinforce rather than diminish said power and privilege, supporting a white saviour complex.

Within academic institutions, related critiques have been made about participatory research (Kesby, 2007; Kindon, 2010; Mohan, 1999; Pain & Francis, 2003). By the early 2000s, shifts in philosophical critique, economic policy and international geopolitics generated a context in which participatory work could "come in from the cold" (Cornwall & Pratt, 2003; Fals-Borda 2006a; Hall, 2005), and it was rapidly promoted in the social and environmental sciences in the Minority World (Brydon-Miller et al., 2004; Jason et al., 2004; Reason & Bradbury, 2008).

But many of the development and research projects that emerged at this time were a far cry from participation's early roots in Freire's emancipatory pedagogy, feminist, or Indigenous practices. Instead, they were examples of tokenistic, illusory, or extractive participatory processes, complicit with what Amy Ritterbusch (Chapter 9) calls the "Nonprofit Industrial Complex" and the "Academic Industrial Complex" (see also People's Knowledge Editorial Collective, 2016). Rather than overturning relationships of power, many actually strengthened them through reauthorising other (non-mainstream or non-Minority World) knowledges as more organic and primitive (Mohan, 1999). In settler-colonial states like Canada, researchers like Geraldine Pratt (2007) wrote about "faux PAR"; while in Australia Jenny Cameron and Kathryn Gibson (2004) warned that even when PAR involved multiple local representations and knowledges, they must be "approached with a degree of caution . . . not blindly accepted at face value as inherently transformative" (p. 8). Within our own engagements with PAR we have grappled with similar constraints and negotiations (Kesby, 2007; Kindon, 2012, 2016a, 2016b; Pain & Francis, 2003).

Clearly, researchers need to be vigilant about their own power and privilege. Unquestioned assumptions about the benevolence of participatory approaches are deeply harmful (Kapoor, 2002; see Cooke & Kothari, 2001). In response to the "tyranny" critique, some have argued that participation is a multifaceted and complex phenomenon, variable across space and time, continually being reinvented, especially through the agency of Majority World communities (see Hickey & Mohan, 2004; Kesby, 2007). Power shifts and morphs beyond what can be predicted, and beyond simplistic notions of good and bad outcomes (Kesby, 2007); "its consequences are not predetermined and its subjects are never completely controlled" (Williams, 2004, p. 557).

As scholars such as Patricia Maguire (1987) and Irene Guijt and Meera Shah (1998) demonstrated, inequalities within communities can also have a strong influence on who does and does not benefit from participatory processes. And, as Table 1.1 notes, feminist critiques have perhaps been most influential in opening up the limitations and dangers of PAR itself, including versions practised within, by, and for communities that are marginalised in relation to centralised power.

Over the last twenty-five years or so, debates around PAR have especially focused on the academy as a toxic context that is hostile to notions of authentic participation. Some academics have been optimistic, citing the agency of external partners: "if we have real respect for the communities we work with, we will understand that they will tell us when we screw up, and they will not let us lead them astray" (Stoecker, 1999, p. 852). Others have been more pessimistic, as Budd Hall (2005) reflected:

> Many of us operate in situations of contradiction and self-conflict . . . is it not possible that in spite of one's personal history, in spite of ideological commitment, in spite of deep personal links with social movements or transformative processes that the structural location of the academy as the preferred location for the organizing of knowledge will distort a participatory research process?
>
> (p. 2)

These debates have been inherently reformist, focused largely on challenges to the practices of Minority World white scholars using PAR within universities. It is hard to ignore that PAR has been slow to alter the fundamental problem of a more powerful external partner wishing to work with communities or elide deeper structural and systemic injustices. While PAR can make inroads and have positive effects, these have tended to be piecemeal and small-scale in practice. Thus, while settler-colonial and neoliberal deployments of participatory research are vital to chart, they are only one aspect of the development of PAR in the Minority World, particularly at a time when the importance of drawing on "epistemologies of the South" is being underscored (Santos, 2016). We now move on to discuss the diversity of critiques and alternative perspectives that largely come from Indigenous and/or racially minoritised researchers within the Minority World.

These have intrinsic value, and offer insights that may help support more critical and critically engaged PAR – an aim of this book.

Critical PAR: Race, Indigeneity and Justice

The People's Knowledge Editorial Collective (2016), for example, have addressed persistent failings in the use of PAR by predominantly white university researchers, where nondominant social groups are still positioned as the subjects of research rather than in control of it; trapped in what the book's subtitle calls a "white-walled labyrinth" (see also People's Knowledge Editorial Collective, 2016). Their criticisms echo earlier statements by Māori scholar Linda Tuhawai Smith (1999, see Table 1.1) in her critique of Western research on Indigenous peoples. Since publication, her foundational book has been used by many Indigenous communities to develop research frameworks that centre Indigenous ownership and control, from the questions asked to the benefits gained (Datta, 2018; Tuck & McKenzie, 2015; see also Kovach, 2009). These frameworks go some way towards the decolonisation of PAR and/or its re-Indigenisation, but care must still be taken with these terms.

North American Indigenous scholars Eve Tuck and Wayne Yang (2016) have constructively taken issue with the proliferation of initiatives claiming to be decolonising (many of which were also participatory). Their landmark paper entitled "Decolonisation is not a Metaphor" (Tuck & Yang, 2016) argued that such initiatives are usually incommensurable with decolonisation. For them, decolonisation fundamentally involves returning lands and other stolen treasures to Indigenous people. However, in practice many decolonising efforts simply fuel settlers' desires to reconcile their guilt and complicity. In so doing, the label *decolonising* frequently "recenters whiteness, it resettles theory, it extends innocence to the settler, it entertains a settler future" (Tuck & Yang, 2012, p3).

Further, and more radically, they extend their critique to Paolo Freire's work on critical consciousness in his *Pedagogy of the Oppressed* (see Table 1.1). For Tuck and Yang (2012), Freire was not sufficiently explicit in his definition of "the oppressed" or what was oppressing them, failing to identify colonisation itself as invasion and theft of lands, but preferring a focus on "internal colonisation" of the mind. His emphasis on the development of critical pedagogies therefore did not offer a reversal of the problem of colonisation, but could be better understood as "settler harm reduction." Tuck and Yang instead advocate for other ways of working by turning to Fanon's (1953) powerful work on colonisation, race, and trauma, and to the construction of freedom and futures in the work of Black feminists such as Lorde (2003).

Their critique, which has been widely influential, has itself seen rejoinders that carefully delineate the differences between, on the one hand, settler-colonial and anti-black projects of decolonisation, and those founded in Indigenous and Black decolonial theory (Curley, Gupta, Lookabaugh, Neubert and Smith, 2022; Garba & Sorentino, 2020). For us, the expansion of Critical PAR in the Minority World is entwined both in Black feminist theory and the development of Indigenous and decolonial research methodologies (Brown & Strega, 2015; Datta, 2018; Lenette, 2022; Tuck & Guishard, 2013). This means that in some situations, PAR can fit well with a decolonial approach to become "an excellent tool to realize the transgressive potential of research that challenges the inherently Western academy because of its roots, values and practices" (Lenette, 2022, p. 27). This fit is contingent and precarious, however, given its frequent appropriation and commodification in neocolonial research (Tuck & Guishard, 2013), and so long as it is used in ways that centre rather than obscure decolonial Majority World and Indigenous cosmologies (Fernández, 2022). Some feminist researchers, including one of us (Kindon, 2012; 2016a), have worked carefully with this in mind, offering a nuanced analysis of power and complicity within a long-term project with Indigenous collaborators, engaging feminist reflexivity in ways that avoid the dangers of self-aggrandisement (Kobayashi, 2003). And Caroline Lenette's (2022) recent book presents a useful rejoinder to the whitewashed literature on PAR, dominated by uncritical Western perspectives.

These critical approaches to PAR frequently return to the original roots and ethical commitments of the approach, founded in activism and the pursuit of systemic change rather than liberal reformism of methodologies and techniques. One powerful example of the development of such Critical PAR (or, as they have termed it, "Liberatory PAR") can be found in the US by scholars at the Public Science Project at City University of New York (e.g., Cahill, Sultana, & Pain, 2007; Cahill, Chapter 2; Fine & Torre, 2019; Stoudt, 2007 and Chapter 5; Torre & Ayala, 2009; Torre, Fine, Stoudt, & Fox, 2012).

In recasting PAR as "public science" (Torre et al., 2012), this group has drawn on critical theory to build on collaborative feminist and antiracist approaches to PAR. They centre US social history and contemporary structural racism within their critique of the research traditions of universities, as well as in fields such as policing, criminal justice, education, and housing that their community research focuses upon, so that:

> PAR is, at once, social movement, social science and a radical challenge to the traditions of science . . . [it] deliberately invert[s] who constructs research questions, designs, methods, interpretations and products; who engages in surveillance. Researchers from the bottom of social hierarchies, the traditional objects of research, reposition as the subjects and architects of critical inquiry, contesting hierarchy and the distribution of resources, opportunities and the right to produce knowledge.
>
> (Fine, Tuck, & Zeller-Berkman, 2007, p. 157)

Concurrently, a critical approach to Youth PAR has developed within the same network of scholars, addressing knotty questions of marginalised young people's participation in and ownership of research when university knowledge and methodologies are dominated by adults (e.g. Cahill, 2004; Cammorota & Fine, 2008; Stoudt, 2007).

Critical Race Theory (CRT) has been a central influence in much Critical PAR work. CRT contends that racism is a key dimension of oppressive structures and institutions (Bell, 1995; Crenshawe, 1991). It draws attention to and challenges the whiteness of educational settings and academic research, the ways that certain knowledges are frequently reproduced, and how certain histories are erased. Following the work of bell hooks (1994) and Freire (1972) on critical pedagogy, CRT emphasises that educational settings and research must be transformed before wider structural change is possible (Torre, 2009). Intersectionality, or how multiple sources of advantage and disadvantage collide and exacerbate each other, is a central concept in CRT (Crenshawe, 1991; see also Lenette, 2022).

CRT also draws attention to the social and political histories of the institutions that have often taken up PAR. In North America and the United Kingdom, for example, many elite universities were founded on slavery, both in terms of finance and slave labour, and rich patrons to this day police the nature of academic knowledge (Chatterjee & Maira, 2014). Although most of these universities today have a web of policies on diversity and inclusion, it is the case that racist, sexist, ableist, homophobic, and transphobic structures and daily practices still discriminate, erase, and exclude certain people from the academy (Ahmed, 2012; Mahtani, 2014). Therefore, attempts by universities or individual academics to engage with communities outside their walls with promises of partnership are often looked upon with suspicion.

For Sara Ahmed (2007), the desire to act and speak for others is symptomatic of white privilege and often, too, of ignorance of the complex politics of social problems. Chatterjee and Maira (2014) argue that the imperial history of Western universities makes the notion of public scholarship deeply problematic: again, is it possible to be radical or insurgent from within? Any collaboration between universities and communities must involve a critique of hegemonic structural relations that is antiracist and anticolonial (Mohanty, 1997).

Maria Torre's (2012) work comparing the principles of CRT and CPAR is helpful here. She suggests that they are both about working "in-between" positions, drawing on Anzaldua's (1987) concept of *nos-otras* that identifies how we (the colonising power) and the other (the colonised group) are so intertwined as to be one system. Torre outlines the shared ethical and political commitments of CPAR and CRT: expanding notions of expert knowledges; recognising that individuals have multiple, overlapping, potentially conflicting identities, loyalties, and allegiances; complicating identity categories; and making the political nature of knowledge construction explicit.

Centrally important for the work in this volume, collaborations then become focused on challenging structural inequality, rather than a donor-recipient white saviour model. As such, the deconstruction of privilege is foundational to and essential in CPAR, if the approach is not to reproduce spaces of injustice.

Having outlined the foundational and more recent theoretical strands informing CPAR in practice, the next section positions this book within this intellectual and political history, outlines the book's aims, and discusses its presences and absences.

Positioning this Book: Presences-Absences

This book has had a long gestation and has altered its form in that time. Our plan several years ago was to create a companion to our 2007 book. That volume provided an introduction to PAR, its politics, practice and methods (Kindon et al., 2007). In this companion volume, we wanted to express our concern about the preponderance of methodological accounts in the literature that seemed to come at the expense of rigorous and nuanced theoretical and political reflections on PAR. We also wanted to move away from the methodological fetishism and celebratory rhetoric that has driven a lot of writing about PAR (at least in the academic journals of Western academia). We wanted to acknowledge scholarship and practice that values and interrogates the critical, political origins of PAR in different parts of the world. Our aim was to tackle and work through confluences and incommensurabilities of different theoretical frames or epistemologies. In developing the vision and proposal for this book, we also wanted to honour our understanding that relations are at the heart of PAR, informed by and entangled within the power geometries of particular institutions, places, and times.

We approached a wide range of scholars across the world and encouraged them to focus on power, tensions of structure-agency, and possibilities (or not) for transformational change and justice at a range of scales. We asked contributors to critically engage with their own practices, and to deconstruct their own privilege within it. This was just before the COVID-19 pandemic broke out.

While our invitations were met with interest and excitement, personal, institutional, and global events impacted the eventual shape of the collection. Colleagues were juggling considerable institutional pressures, employment precarity, ill health, bereavement, and heightened caring responsibilities. Several dropped out as a result of these demands. Significantly, many colleagues' initial commitments to involve their partners as co-authors also became impossible due to the impacts of the COVID-19 pandemic. Research partners, primarily in the Majority World or from Indigenous or minoritised communities in the Minority World, had other priorities of a more urgent nature than writing an academic book chapter – heightened demands to secure their livelihoods and/or care for ailing relatives as the world shifted.

Thus, while the remaining authors integrate many co-researcher and participant views, voices and perspectives into their rich accounts, and acknowledge many who have contributed to their thinking, we recognise the irony that this book, focusing on CPAR, has ended up being written mostly by authors and editors who are located in either the centre of British colonial rule (UK), or formerly British, now settler-colonial societies (US, Australia, and South Africa). We therefore acknowledge the missing perspectives and insights from researchers engaging PAR critically who originate from, or live and work, in other contexts, and we direct readers to other recent publications (see Burns et al., 2021; Lenette, 2022; People's Knowledge Editorial Collective, 2016; Kemmis, McTaggart, & Nixon, 2014) where more diverse voices can be found.

The resulting authors also largely identify as white, but what that means within our different institutional, research, and national contexts varies. We embody a diversity of other markers of social difference associated with gender, sexuality, family status, immigration status, and ability. Some of these dimensions are evident in our writings, others implied or only marginal to discussions. The authors also mostly work within university contexts, as academic researchers, but not all are fulltime, permanent academics, not all are tenured, and not all teach within their roles. One author left academia after completing their chapter here. We write from different disciplinary backgrounds – human geography, urban studies, environmental psychology, migration studies, media and communication studies, and development studies. As such, this book broadly fits within the social sciences and the humanities and lacks perspectives from colleagues evolving and adapting PAR within scientific, education, health, and environmental domains.

Given this genealogy and geopolitics at play in the book's evolution, its aims shifted slightly to:

1. offer readers reflections and insights into the inevitable entanglements of power that come with attempting CPAR in different institutional and geopolitical contexts;
2. critically (re)engage with CPAR from the perspectives of embodied, lived experiences, relationalities and circuits of privilege (Stoudt, Fox, & Fine, 2012);
3. share examples that are critically engaging for those involved and for those reading about them.

This collection is therefore particularly suited to doctoral candidates, early career researchers, and other academic researchers who want to reflect on and innovate in their praxis. However, we hope that it is also helpful to senior undergraduate students learning about participatory approaches, as well as more experienced researchers within and beyond academia. The chapters that follow provide nuanced discussion of their authors' theoretical-ethical framings, navigation and deconstruction of privilege, and the joys and challenges of CPAR.

At this point, readers may be wondering why we have chosen to write in detail about the book's genesis. Academic institutions are notorious for creating artificial divisions between the professional and the personal, between work and life, between forms of productive and reproductive labour. Yet, feminist, Indigenous, and critical race scholars have long called for greater recognition of the messiness and inseparability of these domains (Cho, Banda, Fernandez, & Aronson, 2023; Lyle, Badenhorst, & McLeod, 2020; McAlpin, 2008; Huopaleinen & Satama, 2019). The ongoing global pandemic both exposed this nexus and reinscribed it. And because PAR and CPAR occur in and between the spaces where various structures, systems, and discourses collide, we felt it was important to acknowledge some of the factors that have both constrained and enabled the generation and eventual production of this volume.

Thus, the book we initially envisaged has quite radically changed, and we acknowledge that the result is open to the criticism of (re)centring problematic patterns of authority when it comes to narrating PAR (see Lenette, 2022). We hope, nevertheless, that the book shares some useful reflections by experienced researcher-practitioners, and in developing the themes covered in the next section, demonstrates an ongoing commitment to a reflexive, collaborative politics that works in solidarity with minoritized others towards epistemic, social, and political justice.

Critical Participatory Action Research: Five Key Themes

The resulting eight chapters draw from specific and ongoing long-term PAR collaborations and/or reflections on different PAR projects over a sustained period. Their authors are or have been working in the UK, the US, South Africa, Australia, India, and Colombia over many years, either as residents, neighbours and friends with their co-researchers, or as academics who have engaged in successive collaborations in situ with them. These compelling examples show the ways in which their commitment to critically interrogate their PAR practices has led to innovative and impactful outcomes for those involved, as well as new understandings pertinent to theory-building. The stories they share capture some of the excitement, challenges, and potential of CPAR. They also reveal frustrations, challenges, confusion, uncertainty, and a wide range of emotions that swirl around the negotiation of power and privilege, and the practice of being "*nos-otros*" (Torre, 2009), drawing on Anzaldua, 1987) whatever form that takes in different kinds of projects and places.

Collectively, these authors' accounts offer a critical reappraisal at the same time as charting new directions. For readers, we hope that these honest insights will offer opportunities for self-reflection and considerations of what might be possible in your own contexts. For us, they also bring into sharp focus five key themes which, we suggest, are emblematic of the wider practice of *Critical* Participatory Action Research. These themes are as follows.

Relational Praxis – Caring and Conceiving with, as and for

Firstly, the contributors emphasise the centrality of relational praxis, care ethics, and accountability in their work.[1] Good relationships are central to any practice of PAR: without them there can be no shared feeling, analysis, or responsibility for action, and no mutual accountability around governance and ethics. Without well-functioning relationships, creativity can be stifled, sustainability is hard, and working towards social, environmental, and epistemic justice is not possible.

Therefore, CPAR praxis is built on relationships: decisions, responsibilities, and activities normally undertaken only by the researcher in more conventional research become things that are negotiated, shared, and co-constructed with participants. Precisely because of these negotiations, there are many ways to pursue relational praxis and implement the "right to research" (Appadurai, 2006) for all participants, which includes the right to plan and undertake it, as well as to access and benefit from its results. CPAR may or may not involve external researchers, who may be asked to act as conduits, catalysts, and/or resource contributors to participants' efforts to research themselves, their communities, and the issues that shape their lives. These external researchers will learn and work with them to take action for change. Research design, issue identification, and question formulation emerge in different ways depending on the relations developed between researcher-participants.

Some of our contributors (e.g. Nina Woodrow, Chapter 6) describe projects that were initiated as what Cornwall (2002; 2004) has called somewhat pejoratively "invited spaces" (i.e., with researchers, activists and service professionals coming together to launch initiatives that seek to address a problematic deficit by recruiting participants to explore and imagine solutions). Others describe opening spaces that were less predefined in their focus and were constituted precisely to allow participants space to reflect on and define what constituted the problem. Here they also enabled the grounding of categories of analysis firmly within the neighbourhoods under investigation rather than the scalar units defined by more distant municipal authorities (see Brett Stoudt, Chapter 5).

In other chapters, authors' projects sought to engage with partners and communities already involved in ongoing participatory praxis (see the chapter by Jackie Shaw with Sowmyaa Bharadwaj, Anusha Chandrasekharan and Dheeraj [hereafter Shaw et al.]). This means that participants were in a strong position to take ownership of the new research, and that participatory praxis was sustainable. Whilst the potential for co-optation and domination in ways that reinscribe colonial relations is a salient risk, where researchers invite community members to participate, where researchers engage with already constituted spaces of participation (see Bawaka Country, Chapter 3) are also enmeshed in power (Kesby, 2007). As Nina Woodrow makes clear (in Chapter 6), attention to the relationality of CPAR praxis was crucial on her project, including how parties worked together to ensure priorities were mutually agreed and benefits shared, though conflicts and tensions were not fully resolved.

Many researchers may profess to care about the things and people they research, but the ethics of participatory action research necessitate an attempt to initiate change (Sultana, 2007). In CPAR, the ethics of care are relational. As Caitlin Cahill points out in Chapter 2, unless the *who* and *what* are cared for and held accountable, and the *how* is co-determined and ongoing, then research cannot truly be *in the interests of* those who are presented/represented, no matter how rigorous formal systems of institutional ethical review may be. Like Caitlin, Brett Stoudt (Chapter 5) reports on a long-term project with communities in New York City, which were subject to the slow violence and injustice of discriminatory policing. He illustrates how, by establishing structures and processes like advisory groups, community conversations, co-analysis and multiple forms of dissemination and reporting, it is possible to ensure that research remains accountable to the communities involved (rather than to the interests, priorities, and ethical frames of professional organisations).

The chapter by John Marnell (Chapter 4) aptly illustrates the particular value that theorising has in CPAR. It describes a praxis of theorising about lived realities that is emergent both from experience (that of external- and/or community-researchers) and from contemporary (predominantly Western) queer theory. Developing *new* ways to queer commonsense understandings, Marnell and partners explore the possibility "for things to be otherwise" and for the emergence of entirely new ways of thinking, doing and being. By comparison, the chapter by Bawaka Country (Chapter 3) illustrates the possibility of challenging dominant paradigms and imagining more sustainable futures via mobilising ancient imaginaries. This is not to say that Indigenous, "more-than-human" ontologies are not analogous with new-materialist perspectives in Western social science (which borrow heavily from them), nor, as John points out (citing Gallacher & Gallagher, 2008), that participants can know themselves transparently or are the only source of credible or authentic knowledge. However, the relationships developed by Bawaka Country ensure that contemporary insights do not simply colonise Indigenous imaginaries. Rather, Indigenous Aboriginal imaginaries take prominence and find fuller expression in the group's critical participatory theorising. Neither of these examples is free from tensions, differences of positionality or perspective, or from limitations about what can be mutually understood, shared, or communicated. However, they nicely illustrate the potential for strong and respectful relationships to ensure that these enduring paradoxes can be negotiated (never fully resolved) by partners in praxis, not simply rehearsed after the event in academic accounts.

Embedded Doings – Creating with, as and for

Secondly, each author evidences their development and nurturing of these relationships over the long-term – far beyond the usual confines of short-termist research – with people who are often marginalised, oppressed or victimised by the state and other institutions. Their community collaborations tackle complex issues of social and environmental injustice through embedded and contextually-appropriate research, consistently adapting and evolving the research form and approach as the collectively-negotiated agendas, knowledges, and needs shift. Some authors have also developed relationships with NGOs and international organisations that impact these communities' lives and livelihoods, where their goals and values align. The long-term nature of these examples of CPAR means that knowledge, skills, and impacts are built up and strengthened cumulatively, growing a capacity for future work.

Participatory research is well known for championing innovative and alternative methods, but the examples of co-creation in the chapters that follow go far beyond the capture of data or generation of disseminable products. The processes of collective storytelling, collaborative script development, and participatory filmmaking in the project described by Nina Woodrow (Chapter 6) enabled participants, working with researchers, artists, activists and educators, to explore radically different experiences and develop a shared cosmopolitan identity and mutual empathy. Moreover, their co-created art and media products communicated participants' voices in ways that try to avoid the problems of translation and ventriloquism often found in conventional research reporting.

John Marnell, in Chapter 4, demonstrates how CPAR arts-based research can offer creative forms of resistance, even in contexts in which it is physically dangerous for marginalised people to speak their name, share their stories, or express their identities and desires. Rather than remaining obscured or being visible only via the stereotypes projected by those who aggressively police normalisation, participants created distinctly queer "zines" that celebrate the breadth and complexity of LGBTQ+ migrants' lives on their own terms. The CPAR approach, and the relational dynamics of the group of which John was a part, meant that they were able to move beyond the more conventional kind of outputs he had originally imagined, and instead collectively produce art and research that were richer and more challenging to dominant representations of queerness.

The fact that critical participatory arts-based work is always far more than just art is well demonstrated in Chapter 3 by Bawaka Country. Drawing on Indigenous media, such as songspirals, is a conscious political act central to the action outcomes of the research. For communities that have been devastated by dispossession of land and through the genocide of populations and their cultures, the practices of intergenerational knowledge sharing and of remaking and celebrating artistic/cultural mores is central to their ability to make claims, to resist continuing postcolonial injustice, and to mould futures in which humans co-become more sustainably with the environment and other beings.

Both Brett Stoudt in Chapter 5 and Caitlin Cahill in Chapter 2 offer a reminder that CPAR need not be an exclusively qualitative or arts-based activity. Instead, they show how qualitative and quantitative methods can work hand in hand in CPAR, and that science is a creative process whose products require no less interpretation than artworks. Moreover, they show that when communities experiencing oppression have access to "the master's tools," they can help "dismantle the master's house" (c.f. Lorde, 2003). Both Caitlin and Brett take seriously the insight underlying Lorde's famous injunction, which is that the master's tools function to erase difference and normalise inequality and injustice. They redeploy them through CPAR collaborations to effect change. For example, in New York City, Brett and partners use the powerful technologies of statistics and mapping to make visible the daily experiences that participants (and the broader community) know intimately to be real, but which were obscured by the way data were usually collated and presented to the public by the police department. Rescaling the spatial units, and visually presenting the statistical data on stops, searches, and arrests in timelapse presentations (including how many of these stops turned out to be unjustified) powerfully demonstrated the uneven geography of relentless, racialised slow violence committed by what Caitlin describes as the digitising, metric-driven carceral state. Further, Caitlin suggests that by flipping the script on the state, and asking what alternative wellbeing-oriented indicators might be used to track the progress of state activity in neighbourhoods, participants rescaled the problem away from pathologised individuals and towards pathologising state mechanisms.

Deconstructing Privilege – Sharing Power and Responsibility with, as and for

Thirdly, the contributors attempt to engage with privilege. Such a focus is especially key for this collection, given the positionalities of the final list of contributors as discussed above. Several of our authors demonstrate how they practise unlearning and hyper-self-reflexivity as academic researchers imbricated in various circuits of advantage, solidarity, and dispossession (Kapoor, 2005; Kindon, 2012; Stoudt et al., 2012).

As we have outlined earlier in this introduction, CPAR is founded on critiques around the power relations of knowledge production. Chapters 5 and 2, by Brett Stoudt and Caitlin Cahill, are again very instructive in this regard. Caitlin urges researchers to engage critically with ethics and avoid conducting research in ways that only reinscribe the violence of austerity, state abandonment, and racial profiling. Brett's simultaneous use and interrogation of quantitative research methods exposes the assumptions and prejudices that can hide within supposedly neutral research tools, seemingly objective data, and established spatial or social categories. By opening access to the tools of academic and state power-knowledge to the perspectives of the communities that academics and states often fail to represent, Brett destabilises the privileged position such tools usually occupy when controlled only by "the experts." He and other participants on the project undermine the unequal and unjust power effects such research tools often produce and legitimise.

The authors of chapters in this volume are reflexive on issues of privilege, and alive to the issue that CPAR is itself enmeshed with power – even as they struggle to deploy the power effects of participatory approaches in ways that are less dangerous than those powers they resist. Caitlin Cahill, for example, refuses to allow the ethical procedures of the institution that employs her to simply assume a privileged insight on what constitutes ethics, without first calling to account that institution's own situated position within settler-colonial histories of racialised structural violence. She is acutely aware of academic predecessors who used marginalised communities as laboratories *for*, rather than partners *in* social analysis. The challenge, as Bawaka Country articulate, is to find ways to engage and account for historic injustices and privileges without becoming paralysed by them or allowing past hurts to prevent the development of present or future mutually beneficial relationships through research. Both Caitlin and Bawaka Country acknowledge researchers' ongoing privilege, and value participants' notions of care and kinship, central to the relationships and ethics developed in their projects. In Caitlin's project, these relations and ways of working enabled the research collective to overcome a "cruel optimism" (Berlant, 2011) that lay hidden in the way they had initially imagined that evidence would lead to change. This allowed them to positively reassess the project's ethics and radicalise its methodology in response to insights on injustice that emerged from the Black Lives Matter movement.

The chapter by Amy Ritterbusch (Chapter 9) celebrates the joy of being able to deploy CPAR approaches to disrupt and transform the privileged spaces of the academy, working with community partners and activists to "bring 'the streets' on to the campus and into the classroom" (see also Kye Askins, Chapter 7). However, such approaches can sometimes unthinkingly burden the communities with whom academics work, for example in the time, resources, and other expectations around their participation. Such a burden can result from a desire to meet some gold standard of participation, or sense of responsibility to facilitate others' empowerment to assuage one's own guilt or complicity in hierarchical and exploitative systems (Kapoor, 2005). Kye Askins, in Chapter 7, reflects that such desires can emerge from academic researchers' privileged perspectives on the research process, and that researchers must instead learn to listen more carefully to what participants or co-researchers wish to get from engagement and what they have the capacity to undertake. Research partnerships can only be truly participatory if all collaborators can refuse or at least negotiate the terms of their involvement (Spathopoulou and Meier, 2023).

In Chapter 9, Amy Ritterbusch also reflects on how development and research work (including CPAR) can get entangled in industrialised forms of extraction, production, and commercialisation in ways that "rampage" through the lives of poor and marginalised communities.

Notwithstanding that both external researchers and participants are frequently ensnared by the same neoliberal logics, they are far from equally positioned in relation to them. External researchers rarely risk as much as their partners, participants or community activists and rarely give as much relative to the personal rewards gained. Amy reflects on the financial improbity of certain research partners and the condemnation they received from their activist peers. From her position of structural privilege, she recognises how much the development and "research industrial complex" demands from volunteer partners who, in their own lives, may struggle to make ends meet. Consequently, she encourages the street activists with whom she has worked to be more forgiving of themselves and others, and to recognise that such breakdowns are primarily the structural outcome of inadequately costed community interventions, not of problematic individuals. In this regard, her work echoes the inventions of Brett and Caitlin above in rescaling the problem away from pathologised individuals and towards unjust funding mechanisms and wider, longstanding structural inequalities.

Feeling/Knowing – Becoming with, as and for

Fourthly, the chapters recognise the integral role of emotions and embodied practices for the generation of knowledges, solidarity, and impact in CPAR. In our earlier section on the histories of PAR, we introduced the Latin American concept of *sentipensante* (Fals-Borda 2001; Rendon, 2014), and the necessary rejection within CPAR of the Western oppositional binary between reason and emotion, acknowledging that head and heart not only work together but that their influence is inseparable.

Many of the authors reflect on the ways in which emotions inspire, energise, and guide the work in which they are engaged. Amy Ritterbusch (Chapter 9), Caitlin Cahill (Chapter 2), and Brett Stoudt (Chapter 5) explain how their work is driven by the grief, righteous anger, and outrage that emerges from participants' visceral experiences of injustice. However, it is frequently bonds of care that ultimately link researchers, participants, and communities as they work together to take action for change. John Marnell (Chapter 4), Amy Ritterbusch (Chapter 9) and Kye Askins (Chapter 7), for example, explore in more detail what Askins calls the "feeling methodologies" of CPAR, and the need to recognise the impacts of investigating issues that involve oppression, violence, and trauma. John suggests that intimacy and love are both a political and ethical imperative; he sees them as essential to the quality and trusting relationships that are central to participatory endeavours, and as "transform[ing] . . . research encounters into sites of radical connection." (Chapter 4) It is through such deep mutual engagement, Kye Askins (Chapter 7) and Bawaka Country (Chapter 3) suggest, that participatory research morphs into a co-becoming of people and place.

Kye Askins in Chapter 7 richly explores the requirement for simultaneous and mutually reinforcing feeling and thinking in CPAR praxis, reflecting that it was participants who taught her to recognise the ontological and epistemological centrality of emotional intelligence to everyday- and research-life. Recognising and holding onto emotions enhances rather than negates the capacity for rational, critical thought. The co-production and co-expression of embodied feelings (be they painful or joyful) is critical to the co-production of knowledge capable of driving progressive change. Like John, Kye asserts that, "'feeling together' holds the potential to open up radical spaces of hope and transformation." Finally, in Chapter 9, Amy Ritterbusch urges that we honour the best memories of the people and movements we work with, forgive present failings, and dream forward to imagined worlds beyond. While we hold each other accountable, relationships of friendship, love and respect can compel us to do this in ways that are empathetic and generous.

Sustaining Actions – within and beyond Research Projects

All the contributors to this volume are engaged in projects that are as much about action to bring about greater equity and justice as they are about researching and making injustice visible. They describe how CPAR relationships and praxis enable fruitful co-investigation and generate spaces that enable participants and researchers to transform their understandings of themselves and others. Crucially however, they also explore how their CPAR projects attempt to impact and transform the world beyond the lives of those who participate in them.

Again, the emotional and affective dimensions of what motivates and sustains action should not be underestimated. Citing J. K. Gibson-Graham (2006), Amy Ritterbusch emphasises that a postcapitalist politics requires new structures of feeling that are more certain in their sense of hope. Nina Woodrow, among others, explores how participants' personal stories and narratives can be shared with wider publics in engaging, participatory ways that connect people, and evoke empathetic responses that provide an impulse for change. One example in her work is an intentionally unfinished short film that simulated various problematic or positive scenarios, instigating audience interaction and engagement at public screening events. This promoted among audiences the kind of self-reflexivity and contemplation of alternative responses that the project participants had already undergone during the development of the film. In her chapter, Kye Askins speaks of the cumulative emotive experiences and emotional learning accruing through engagement with CPAR, positively informing her own life and work and that of her participants and students, as each continues to resist social injustice beyond specific projects.

Bawaka Country propose that their inspiring accounts of weaving, painting, and songspirals should not be taken out of context as merely cultural artefacts to be admired (or worse, collected), since they are simultaneously also enactments of law and claims to sovereignty. However, they may nevertheless inspire other folk to listen harder to the environment and history of their own contexts to find ways to live with other humans and nonhumans that are more artful, harmonious, and regenerative.

Both Brett Stoudt and Caitlin Cahill demonstrate how the power-knowledges that produce everyday experiences of injustice can be engaged and challenged in ways that ensure that the effects of co-produced research can roll out beyond the immediate confines of projects and provide resources for continued action and challenges to power. John Marnell suggests that the action inherent in CPAR need not be limited to economic and material change, such as instigating improved public access to services or to challenging and reforming state institutions. Action also consists of the disruption and queering of social categories that allow for the possibility of thinking, being, and doing otherwise. Across the chapters, the authors show how several normative categories (e.g., data, scale, identity, centre and margin, lay and expert, and researcher and researched) are reconceptualised as an integral part of CPAR. In an era in which reactionary and revisionist voices loudly demonstrate the powerful effects of narrative, and the oppressive work that categories can do, it has never been more important to both destabilise oppressive norms and to defend more just positions to replace them.

Importantly, Amy Ritterbusch reminds us of the fragility of CPAR and the relationships and relations it facilitates. Initiatives do not always succeed and can crumble under the pressure of attempts to fulfil commitments to partners, communities, and self, in contexts that are contradictory and unequal, notwithstanding attempts to make them otherwise. In such circumstances, Amy urges us to sustain the spirit of CPAR by remembering the best of what was achieved, forgive failures, and dream of better futures.

In this context, Chapter 8 by Jackie Shaw et al., offers perhaps the most in-depth analysis of the challenges of sustaining and distancing the effects of CPAR, through their rich discussions of scaling. Firstly, scaling in spatial terms: *scaling down* decision-making; *scaling out* activities over wider areas, multiple sites and more people *scaling up* local knowledge to influence national and international policy; and *scale jumping* that questions normative scale imaginaries (see Chapter 2 and Chapter 5). Secondly, scaling in temporal terms: *scaling long* through time to avoid short term "single loop problem solving." Jackie and colleagues point out that scaling processes are complex and require careful thought, as well as needing to anticipate and counter the negative impacts of time and resourcing constraints, and co-option by powerful interests, which can dilute the transformative promise of CPAR. They also discuss new challenges presented by scaling out, as the care and intimate relations developed at local scales need to adjust and accommodate wider audiences and publics with perhaps different and contradictory perspectives. Further work, contextual sensitization and (sometimes agonistic) dialogue is necessary if such scaling is to be successful, if local input is to be retained, and if existing hierarchies are not to be reinstated.

As many of the contributors demonstrate, the impacts of participatory research should not be measured against a sliding scale of importance from the local to the global (Pain et al., 2011), partly as their intensities of relation and depth of sustained action are often greatest in local communities. However, the work of Jackie Shaw et al., provides one example where broader national issues were explored by participants locally, and where these local perspectives were brought into and influenced international policy space. As other contributors also demonstrate at other scales, the innovative use of digital technology can enable marginalised people to be seen and heard, including in the global decision-making arenas where other voices normally dominate.

Concluding Remarks

As this chapter has outlined, PAR and CPAR have long and variegated histories. Their origins in, and influences from, the struggles of marginalised and minoritised peoples of the Majority World and those of Indigenous peoples, remind us of the need to foster *sentipensante*, conscientisation, and collective action as *nos-otras* towards more just systems at different scales. Nonetheless, with PAR's diffusion from the Majority to the Minority World and pressure to be seen to engage publics and co-produce knowledge within neoliberal higher education, many iterations of research and action claiming to be participatory are a far cry from those origins and practices. While some might still be judged preferable approaches to colonial, extractive, and potentially exploitative research, some instances reiterate those dominant patterns, and many are depoliticised and lack a critical orientation.

PAR is at a critical juncture, entangled within debates about epistemic justice, decolonisation, and the creation of more just and regenerative futures. In curating this edited collection, our aims have been to revisit and evaluate Critical PAR by: (i) offering reflections and insights into the inevitable entanglements of power that come with attempting CPAR in different institutional and geopolitical contexts; (ii) critically (re)engaging with CPAR from the perspectives of embodied lived experiences, relationalities and circuits of privilege (Stoudt et al., 2012); and (iii) sharing examples that are critically engaging for those involved and for those reading about them. The chapters that follow bring to life the five themes we have outlined above, and that we consider to be central to the practice and future of Critical PAR: relational praxis, embedded doings, deconstructing privilege, feeling/knowing, and sustaining actions.

The discussions in this chapter and those that follow also inform our proposed framework for future action in Table 1.2. Acknowledging our privileged positions and the responsibilities that accompany these, this framework calls academic researchers into active reform and decolonisation of our institutions, including ethics and funding mechanisms. In CPAR partnerships, it calls for a renewed commitment to frame praxis in ways that honour and benefit from diverse epistemologies, co-becoming with place, and solidarities across scale. Finally, we advocate for relational responsibility which goes beyond power-sharing to having skin in the game.

By sharing reflexive accounts of CPAR in this volume, we call on scholars and activists to be bold and explicit about the politics of their work, to draw on CPAR's historical antecedents and to reclaim the critical dimensions of participatory praxis in their future endeavours. Such work will involve consciously engaging the risks and constraints associated with epistemological incommensurabilities and structural inequalities, as well as commitment, creativity, and compassion. There is still much to do to make injustices visible, and to address violent continuities of colonialism and capitalism.

We thank our colleagues for their contributions to this volume, participatory scholarship, and the wider world we inhabit, and we hope that readers find this book inspiring and supportive for their own work and efforts.

In solidarity,
Sara, Rachel, and Mike

Table 1.2 Engaging PAR Critically: Actions for Academic Researchers

Radical reform	*Regular reframing*
• Reflexively attend to researcher privilege and power • Work within institutions and across scales to challenge the neoliberalisation and marketisation that has led to depoliticised deployments of PAR • Advocate for better resourcing and recognition of CPAR, and investment in long term collaborations • Support decolonising research and enhance epistemic justice	• Trouble what is normative to identify and challenge privilege, assumptions, limiting categories, and the work they do in perpetuating inequality and injustice • Support calls to Indigenise and queer research while attending to potential intersecting hierarchies • Recognise the specificity of place in the co-construction of embedded, emotional, and embodied knowledges • Promote interscalar alliances
Relational responsibility	
• Negotiate power, knowledge and co-becoming to foster solidarity and action • If in positions of privilege, identify where we have skin in the game so as to listen deeply and navigate *nos-otros* • Embrace the improvised uncertainties and the risks of being accomplices in forging a better world in challenging times	

Note

1 By including "as" in this section heading and with other themes, we are acknowledging that CPAR partners become "enmeshed in each other's lives" (Guishard, 2015 p. iv) generating knowledge intersubjectively.

References

Ahmed, S. (2012). *On Being Included: Racism and Diversity in Institutional Life*. Durham, NC: Duke University Press.

Alexander, C., Beale, N., Kesby, M., Kindon, S., McMillan, J., Pain, R., & Ziegler, F. (2007). Participatory diagramming: A critical view from North East England. In S. Kindon, R. Pain, & M. Kesby (Eds.), *Participatory Action Research Approaches and Methods: Connecting People, Participation and Place*. 112–121. New York: Routledge.

Anzaldua, G. (1987) *Borderlands/La frontera*. San Francisco, CA: Aunt Lute Books.

Appadurai, A. (2006). The right to research. *Globalization, Societies and Education* 4(2): 167–177.

Banks, S., Hart, A., Pahl, K., & Ward, P. (Eds.) (2018). *Co-Producing Research: A Community Development Approach*. Bristol: Bristol University Press.

Bell, D. A. (1995). Who's afraid of critical race theory? *University of Illinois Law Review*, 1995(4): 893–910.

Boal, A. (2006). *Theatre of the Oppressed*. London, United Kingdom: Pluto Press.

Boal, A. (2009). *The Aesthetics of the Oppressed*. London: Routledge.

Brown, L., & Strega, S. (Eds.) (2015). *Research as Resistance: Critical, Indigenous, and Anti-Oppressive Approaches*. Toronto, Canada: Canadian Scholars Press.

Brydon-Miller, M., Maguire, P., & McIntyre, A. (Eds.). (2004). *Travelling Companions: Feminism, Teaching and Action Research*. Westport, Connecticut: Praeger.

Burns, D., Howard, J., & Ospina, S. (Eds.) (2021). *The SAGE Handbook of Participatory Research and Inquiry*. London: Sage Publications.

Burton, S., & Bowman, B. (2022). The academic precariat: Understanding life and labour in the neoliberal academy. *British Journal of Sociology of Education*, 43(4): 497–512. https://doi.org/10.1080/01425692.2022.2076387.

Cahill, C. (2004). Defying gravity? Raising consciousness through collective research. *Children's Geographies*, 2(2): 273–286.

Cahill, C., Sultana, F., & Pain, R. (2007). Participatory ethics: Politics, practices, institutions. *ACME: An International Journal for Critical Geographies*, 6(3): 304–318. https://www.acme-journal.org/.

Cameron, J., & Gibson, K. (2005). Participatory action research in a poststructuralist vein. *Geoforum*, 36(3): 315–331.

Cammorota, J., & Fine, M. (2008). Youth participatory action research: A pedagogy for transformational resistance. In J. Cammarota & M. Fine (Eds.), *Revolutionizing Education: Youth Participatory Action Research in Motion*. 1–11. New York, Routledge.

Chambers, R. (1983). *Rural Development: Putting the Last First*. London: Longman.

Chambers, R. (1994). *Paradigm Shifts and the Practice of Participatory Research and Development*. IDS Working Paper 2. Brighton: Institute of Development Studies.

Chambers, R. (1997). *Whose Reality Counts? Putting the First Last*. Rugby: Practical Action Publishing.

Chambers, R. (2002). *Participatory Workshops: A Sourcebook of 21 Sets of Ideas and Activities*. London: Routledge.

Chatterjee, P., & Maira, S. (Eds.) (2014). *The Imperial University: Academic Repression and Scholarly Dissent*. Minnesota: University of Minnesota Press,.

Cho, K., Banda, R., Fernandez, E., & Aronson, B. (2023). Testimonios de las atravesadas: A borderland existence of women of color faculty. *Gender, Work and Organization*, 30(2): 724–743. https://doi.org/10.1111/gwao.12894.

Cleaver, F. (1999). Paradoxes of participation: Questioning participatory approaches to development. *Journal of International Development*, 11: 597–612. https://doi.org/10.1002/(SICI)1099-1328(199906)11:4<597:AID-JID610>3.0.CO;2-Q.

Cooke, B., & Kothari, U. (2001). *Participation: The New Tyranny?* London: Zed Books.

Coombes, B., Johnson, J., & Howitt, S. (2014). Indigenous Geographies III: Methodological Innovation and the unsettling of participatory research. *Progress in Human Geography*, 38(6): 845–854. https://doi.org/10.1177/0309132513514723.

Coombes. B. (2017). Kaupapa Māori research as participatory enquiry: Where's the action? In B. Coombes, T. K. Hoskins, & A. Jones (Eds.), *Critical Conversations in Kaupapa Māori*. 29–48. Wellington, Huia Publishers.

Cornwall, A. (2002). *Making Spaces, Changing Places: Situation Participation in Development*. IDS Working Paper, 170. Brighton: Institute of Development Studies. https://www.ids.ac.uk/publications/making-spaces-changing-places-situating-participation-in-development/.

Cornwall, A. (2004). Spaces for transformation? Reflections on issues of power and difference in participation in development. In S. Hickey & G. Mohan (Eds.), *Participation: From Tyranny to Transformation*. 75–91. London: Zed Books.

Cornwall, A. (2017). Introduction: New democratic spaces? The politics and dynamics of institutionalised participation. *IDS Bulletin*, 48(1A): 1–10. https://doi.org/10.19088/1968-2017.144.

Cornwall, A., & Pratt, G. (2003). The trouble with PRA: Reflections on dilemmas of quality. *PLA Notes*, 47: 38–44.

Crenshaw, K. (1991). Mapping the margins: Intersectionality, identity politics, and violence against women of color. *Stanford Law Review*, 43(6): 1241–1299. https://doi.org/10.2307/1229039.

Curley, A., Gupta, P., Lookabaugh, L., Neubert, C., & Smith, S. (2022). Decolonisation is a political project: Overcoming impasses between Indigenous sovereignty and abolition. *Antipode*, 54 (4): 1043–1062. https://doi.org/10.1111/anti.12830.

Datta, R. (2018). Decolonizing both researcher and research and its effectiveness in Indigenous research. *Research Ethics*, 14(2): 1–24. https://doi.org/10.1177/1747016117733296.

de Sousa Santos, B. (2014). Epistemologies of the south and the future. *From the European South*, 1: 17–29.

Fals-Borda, O. (1995). *Research for social justice: Some North-South convergences*. Plenary address at the Southern Sociological Society Meeting, Atlanta, GA. http://comm-org.wisc.edu/si/falsborda.htm.

Fals Borda, O. (2001). Participatory (action) research in social theory: Origins and challenges. In P. Reason & H. Bradbury (Eds.), *Handbook of Action Research: Participative Inquiry and Practice*. 27–37. London: Sage Publications.

Fals-Borda, O. (2006a). P(A)R in social theory: Origins and challenges. In P. Reason & H. Bradbury (Eds.), *Handbook of Action Research*. 27–37. London: Sage Publications.

Fals-Borda, O. (2006b). The North-South convergence: A 30-year first-person assessment of PAR. *Action Research*, 4(3): 351–358. https://doi.org/10.1177/1476750306066806.

Fanon, F. (1953). *Black Skin, White Masks*. London: Grove Press.

Fernández, J. S. (2022). Decolonising participatory action research in community psychology. In S. Kessi, S. Suffla, & M. Seedat (Eds.), *Decolonial Enactments in Community Psychology*. 29–51. Cham, Switzerland: Springer.

Fine, M., Tuck, J. E., & Zeller-Berkman, S. (2007). Do you believe in Geneva? In N. Denzin, L. T. Smith, & Y. Lincoln (Eds.), *Handbook of Critical and Indigenous Knowledges*. 157–180. Beverley Hills, CA: Sage Publications.

Fine, M., & Torre, M. E. (2019). Critical participatory action research: A feminist project for validity and solidarity. *Psychology of Women Quarterly*, 43(4): 433–444. https://doi.org/10.1177/0361684319865255.

Fisher, P. A., & Ball, T. J. (2003). Tribal participatory research: Mechanisms of a collaborative model. *American Journal of Community Psychology, 32*(3–4): 207–216. https://doi.org/10.1023/B:AJCP.0000004742.39858.c5.

Freire, P. (2007). *Pedagogy of the Oppressed*. New York: Continuum.

Gandhi, M. K. (1928). *Satyagraha in South Africa*. Ahmedabad, India: Navajivan.
Garba, T., & Sorentino, S-M. (2020). Slavery is a metaphor: A critical commentary on Eve Tuck and K. Wayne Yang's "Decolonization is not a metaphor." *Antipode*, 52(3): 764–782.
Gayá, P. (2021). Towards ever more extended epistemologies: Pluriversity and decolonisation of knowledge in participatory inquiry. In D. Burns, J. Howard, & S. Ospania (Eds.), *The Sage Handbook of Participatory Research and Inquiry*. 169–184. London: Sage.
Gergen, K. (2015). Culturally inclusive psychology from a constructionist standpoint. *Journal for the Theory of Social Behaviour*, 45(1): 95–107.
Glassman, M., & Erdem, G. (2014). Participatory action research and its meanings: Vivencia, praxis, conscientization. *Adult Education Quarterly*, 64(3): 206–221. https://doi.org/10.1177/0741713614523667.
Guijt, I. & Shah, M. (1998). *The Myth of Community: Gender Issues in Participatory Development*. Rugby: Intermediate Technology Publications.
Guishard, M. A. (2015). *Nepantla and Ubuntu Ethics Para Nosotros: Beyond Scrupulous Adherence toward Threshold Perspectives of Participatory/Collaborative Research Ethics*. PhD thesis. New York:City University of New York. https://academicworks.cuny.edu/gc_etds/957/.
Hall, B. (2005). In from the cold? Reflections on participatory research from 1970–2005. *Convergence*, 38(1): 5–24. https://chairerp.uqam.ca/fichier/document/Publications/In_From_the_Cold_Reflections_on_Participatory_Research_from_1970-2005.pdf.
Heron, J., & Reason, P. (1997). A participatory inquiry paradigm. *Qualitative Inquiry*, 3(3): 274–294.
Hickey, S., & Mohan, G. (2004). *Participation: From Tyranny to Transformation*. London: Zed Books.
hooks, b. (1994). *Teaching to Transgress*. London: Routledge.
Huopalainen, A., & Satama, S. (2019). Mothers and researchers in the making: Negotiating 'new' motherhood within the 'new' academia, *Human Relations*, 71(1): 98–121. https://doi.org/10.1177/001872671876457.
Jason, L. A., Keys, C. B., Suarez-Balcazar, Y., Taylor, R. R., Davis, M., Durlak, J., & Isenberg, D. (Eds.). (2004). *Participatory Community Research: Theories and Methods in Action*. Washington, DC: American Psychological Association.
Kanyamuna, V., & Zulu, K. (2022). Participatory research methods: Importance and limitations of participation in development practice. *World Journal of Social Sciences and Humanities*, 8(1): 9–13. http://pubs.sciepub.com/wjssh/8/1/2.
Kapoor, I. (2002). The devil's in the theory: A critical assessment of Robert Chambers' work on participatory development. *Third World Quarterly*, 23(1): 101–117. https://doi.org/10.1080/01436590220108199.
Kapoor, I. (2005). Participatory development, complicity and desire. *Third World Quarterly*, 26(8): 1203–1220. http://hdl.handle.net/10315/7851.
Kemmis, S., McTaggart, R., & Nixon, R. (Eds.) (2014). *The Action Research Planner: Doing Critical Participatory Action Research*. London: Springer.
Kesby, M. (2000). Participatory diagramming: Deploying qualitative methods through an action research epistemology. *Area 34*, (4): 423–435.
Kesby, M. (2007). Spatialising participatory approaches: The contribution of geography to a mature debate. *Environment and Planning A*, 39(12): 2813–2831. https://doi.org/10.1068/a38326.
Kesby, M., Kindon, S., & Pain, R. (2005). Participatory approaches and diagramming techniques. In R. Flowerdew & D. Martin (Eds.), *Methods in Human Geography: A Guide for Students Doing a Research Project*. 144–166. Harlow: Pearson Education.
Kesby, M., Kindon, S., & Pain, R. (2007). Participation as a form of power: Retheorising empowerment and spatialising participatory action research. In S. Kindon, R. Pain, & M. Kesby (Eds.), *Participatory Action Research Approaches and Methods: Connecting People, Participation and Place*. 19–25. New York: Routledge.
Kindon, S. (1995). Dynamics of difference: Exploring empowerment methodologies with women and men in Bali. *New Zealand Geographer*, 51(1): 10–12.

Kindon, S. (2003). Participatory video in geographic research: A feminist practice of looking? *Area*, 35(2): 142–153. http://www.jstor.org/stable/20004304.

Kindon, S. (2010). Participation. In S. Smith, R. Pain, S. Marston & J. P. Jones III (Eds.), *The SAGE Handbook of Social Geographies*. 517–545. London: Sage.

Kindon, S. (2012). *Thinking-through-Complicity with Te Iwi o Ngāti Hauiti: Towards a Critical Use of Participatory Video Research*. PhD thesis. Hamilton: University of Waikato.

Kindon, S. (2016a). Participatory video as a feminist practice of looking: 'Take two!' *Area*, 48(4): 496–503. https://doi.org/10.1111/area.12246.

Kindon, S. (2016b). Participatory video's spectro-geographies. *Area*, 48(4): 449–451. https://doi.org/10.1111/area.12215.

Kindon, S., & Latham, A. (2002). From mitigation to negotiation: Ethics and the geographic imagination. *New Zealand Geographer*, 58(1): 14–22.

Kindon, S., Pain, R., & Kesby, M. (2007). Participatory action research: Origins, approaches and methods. In S. Kindon, R. Pain, & M. Kesby (Eds.), *Participatory Action Research Approaches and Methods: Connecting People, Participation and Place*. 9–18. New York: Routledge.

Kobayashi, A. (2003). GPC ten years on: Is self-reflexivity enough? *Gender, Place and Culture: Journal of Feminist Geography*, 10(4): 345–349.

Kothari, U. (2001). Power, knowledge and social control in participatory development. In B. Cook & U. Kothari (Eds.), *Participation the New Tyranny?* 139–152. London: Zed Books.

Kothari, U. (2005). Authority and expertise: The professionalisation of international development and the ordering of dissent. *Antipode*, 37(3): 425–446.

Kovach, M. E. (2009). *Indigenous Methodologies: Characteristics, Conversations and Contexts*. Toronto: University of Toronto Press.

Kovach, M. (2015). Emerging from the margins: Indigenous methodologies. In L. Brown & S. Strega (Eds.), *Research as Resistance: Critical, Indigenous, and Anti-Oppressive Approaches*. 43–64. Toronto: Canadian Scholars Press.

Leal, P. (2007). Participation: The ascendancy of a buzzword in the neo-liberal era. *Development in Practice*, 17(4–5): 539–548. https://doi.org/10.1080/09614520701469518.

Lenette, C. (2022). *Participatory Action Research: Ethics and Decolonization*. Oxford: Oxford University Press.

Lewin, K. (1946). Action research and minority problems. *Journal of Social Issues*, 2(4): 34–46.

Lorde, A. (2003). The master's tools will never dismantle the master's house. In R. Lewis & S. Mills (Eds.), *Feminist Postcolonial Theory: A Reader*. 25–27. New York: Routledge.

Lyle, E., Badenhorst, C., & McLeod, H. (2020). Archives, aesthetic dimensions, and academic identity. *Canadian Review of Art Education*, 47(1): 7–21.

Maguire, P. (1987). *Doing Participatory Research: A Feminist Approach*. Amherst, MA: The Centre for International Education, University of Massachusetts.

Mahtani, M. (2014). Toxic geographies: Absences in critical race thought and practice in social and cultural geography. *Social & Cultural Geography, 15*(4): 359–367.

Mason, O., & Megoran, N. (2021). Precarity and dehumanization in higher education. *Learning and Teaching*, 14(1): 35–59. https://doi.org/10.3167/latiss.2021.140103.

McAlpin, J. (2008). *Place and Being: Higher Education as a Site for Creating Biskabii – Geographies of Indigenous Academic Identity*. PhD thesis. Urbana, IL: University of Illinois at Urbana-Champaign.

McKittrick, K. (2021). *Dear Science and Other Stories*. Durham, NC: Duke University Press.

Mohan, G. (1999). Not so distant, not so strange: The personal and the political in participatory research. *Ethics Place and Environment*, 2(1): 41–54. https://doi.org/10.1080/13668799908573654.

Mohanty, C. (1997). Under western eyes: Feminist scholarship and colonial discourses. In A. McClintok, A. Mufti, & E. Shohat (Eds.), *Dangerous Liaisons: Gender, Nation and Postcolonial Perspectives*. 255–277. Minnesota: University of Minnesota Press.

Monk, J., Manning, P., & Denman, C. (2003). Working together: Feminist perspectives on collaborative research and action. *ACME: An International E-Journal for Critical Human Geographies*, 2(1): 91–106. https://acme-journal.org/index.php/acme/article/view/710.

Morley, L. (2015). Troubling intra-actions: Gender, neoliberalism and research in the global academy. *Journal of Education Policy*, 31(1): 28–45. https://doi.org/10.1080/02680939.2015.1062919.

Moyo, L. (2020). Decolonial research methodologies: Resistance and liberatory approaches. In L. Moyo, *The Decolonial Turn in Media Studies in Africa and the Majority World*. 187–225. Cham: Palgrave Macmillan.

mrs c kinpaisby-hill (2011). Participatory praxis and social justice: Towards more fully social geographies. In V. M. Del Casino, M. Thomas, P. Cloke & R. Panelli (Eds.), *A Companion to Social Geography*. 214–234. London: Wiley Blackwell.

O'Keefe, T., & Courtois, A. (2019). 'Not one of the family': Gender and precarious work in the neoliberal university. *Gender, Work and Organization*, 26(4): 463–479 https://doi.org/10.1111/gwao.12346.

Ospina, S., Burns, D., & Howard, J. (2021). Introduction to the handbook: Navigating the complex and dynamic landscape of participatory research and inquiry. In D. Burns, J. Howard, & S. Ospina (Eds.), *The Sage Handbook of Participatory Research and Inquiry*. 3–16. London: Sage Publications.

Pain, R., & Francis, P. (2003). Reflections on participatory research. *Area*, 35(1): 46–54. https://www.jstor.org/stable/20004288.

Pain, R., Kesby, M., & Askins, K. (2011). Geographies of impact: Power, participation and potential. *Area*, 43(2): 183–188. https://doi.org/10.1111/j.1475-4762.2010.00978.x.

Pain, R., Kindon, S., & Kesby, M. (2007). Making a difference to theory, practice and action. In S. Kindon, R. Pain, & M. Kesby (Eds.), *Participatory Action Research Approaches and Methods: Connecting People, Participation and Place*. 52–58. New York: Routledge.

Parker, L., Martin-Sardesai, A., & Guthrie, J. (2023). The commercialized Australian public university: An accountingized transition. *Financial Accountability and Management*, 39(1): 125–150. https://doi.org/10.1111/faam.12310.

People's Knowledge Editorial Collective (2016). *People's Knowledge and Participatory Action Research: Escaping the White-Walled Labyrinth*. Rugby: Practical Action Publishing Ltd.

Pratt, G. (2007). Working with migrant communities: Collaborating with the Kalayaan Centre in Vancouver, Canada. In S. Kindon, R. Pain, & M. Kesby (Eds.), *Participatory Action Research Approaches and Methods: Connecting People, Participation and Place*. 121–129. New York: Routledge.

Rappaport, J. (2020). *Cowards Don't Make History: Orlando Fals Borda and the Origins of Participatory Action Research*. Durham, NC: Duke University Press.

Reason, P., & Bradbury, H. (2008). *The Sage Handbook of Action Research: Participative Inquiry and Practice*. London: Sage Publications.

Rendon, L. I. (2014). *Sentipensante (Sensing/Thinking) Pedagogy*. New York: Stylus Publishing.

Riley, S., & Reason, P. (2015). Co-operative inquiry: An action-research practice. In J. Smith (Ed.), *Qualitative Psychology: A Practical Guide to Research Methods*. 168–198. London: Sage Publications.

Santos, B. (2018). *The End of the Cognitive Empire: The Coming of Age of Epistemologies of the South*. Durham, NC: Duke University Press.

Spathopoulou, A., & Meier, I. (2023). Practising refusal as relating otherwise: Engagements with knowledge production, 'activist' praxis and borders. *Fennia: International Journal of Geography*, 201(2): 140–153. https://doi.org/10.11143/fennia.137167.

Stoecker, R. (1999). Are academics irrelevant? Roles for scholars in participatory research. *American Behavioral Scientist*, 42(5): 840–854. https://doi.org/10.1177/00027649921954561.

Stoudt, B. G. (2007). The co-construction of knowledge in "safe spaces": Reflecting on politics and power in participatory action research. *Children, Youth and Environments*, 17(2): 280–297. https://www.jstor.org/stable/10.7721/chilyoutenvi.17.2.0280.

Stoudt, B., Fox, M., & Fine, M. (2012). Contesting privilege with critical participatory action research. *Journal of Social Issues*, 68(1): 178–193.

Sultana, F. (2007). Reflexivity, positionality and participatory ethics: Negotiating fieldwork dilemmas in international research. *ACME: An International E-Journal for Critical Geographies*, 6(3): 374–385. https://acme-journal.org/index.php/acme/article/view/786.

Swantz, M. L. (1985). *Women in Development: A Creative Role Denied? The Case of Tanzania.* London: C. Hurst / St Martin's Press.

Torre, M. E. (2009). Participatory action research and critical race theory: Fuelling spaces for nos-otras to research. *The Urban Review*, 41: 106–120. https://doi.org/10.1007/s11256-008-0097-7.

Torre, M. E., & Ayala, J. (2009). Envisioning participatory action research entremundos. *Feminism & Psychology*, 19(3): 387–393. https://doi.org/10.1177%2F0959353509105630.

Torre, M. E., Fine, M., Stoudt, B., & Fox, M. (2012). Critical participatory action research as public science. In P. Camic & H. Cooper (Eds.), *The Handbook of Qualitative Research in Psychology: Expanding Perspectives in Methodology and Design.* 171–184. Washington, DC: American Psychological Association.

Tuck, E., & Guishard, M. (2013). Uncollapsing ethics: Racialized sciencism, settler coloniality, and an ethical framework of decolonial participatory action research. In T. Kress, C. Malott, & B. Porfilio (Eds.), *Challenging Status Quo Retrenchment: New Directions in Critical Research.* 3–27. Charlotte, NC: Information Age Publishing.

Tuck, E., & McKenzie, M. (2015). *Place in Research: Theory, Methodology and Methods.* London: Routledge.

Tuck, E., & Yang, K. W. (2012). Decolonization is not a metaphor. *Decolonization: Indigeneity, Education & Society*, 1(1): 1–40. https://jps.library.utoronto.ca/index.php/des.

Tuck, E., & Yang. K. W. (2016). What justice wants. *Critical Ethnic Studies*, 2(2): 1–15. https://doi.org/10.5749/jcritethnstud.2.2.0001.

Tuhiwai Smith, L. T. (1999). *Decolonizing Methodologies: Research and Indigenous Peoples.* London: Zed Books.

Whyte, W. F., Greenwood, D. J., & Lazes, P. (1989). Participatory action research: Through practice to science in social research. *American Behavioral Scientist*, 32(5): 513–551.

Williams, G. (2004). Evaluating participatory development: Tyranny, power and (re)politicisation. *Third World Studies Quarterly*, 25(3): 557–578. https://doi.org/10.1080/0143659042000191438.

2 "Can We Track Human Dignity?"

Critical Participatory Ethics and Care

Caitlin Cahill

Figure 2.1 Film still, Stephfon, video testimonial, GrowingUpPoliced.org

> I was coming home with some friends and the cops had stopped us, and they jumped out the car – and they were, like, told us to put our hands against the wall. And we were like, "What did we do?" and they were just like, "Shut up!" They started patting us down, and I don't really know my rights that well so I didn't know what to say. So, I just did what they told me to do basically. And they checked us down and then I'm like, "Oh, are you finished and stuff? Can we go?" And they were like "Give me your ID [identification]," but at that time I didn't have an ID, so then, he's like, "You know you could go to jail for not having an ID" and I was, like "We didn't do nothing wrong, we were just walking." So, then he's, like oh, "Just be quiet!" – so I just shut up. And that's when he ran our names, and that's when he saw we had no warrants, and that's when he was like, "Alright y'all free to go." And then as we were walking away, he said "I better not catch you guys over here again."
>
> (Stephfon, 17, Brooklyn)

Bravely sharing his experience with the police in a public testimonial, Stephfon creates a social and shared context for witnessing his encounter. This is what he wanted to do as part of his activism and participatory action research: to speak to you – and to all of us – to challenge the dominant narrative about policing and the criminalisation of communities of colour. Stephfon was one of millions of young people of colour, mostly Black and Latino young men, who were stopped by the police after 9/11, when the police presence was dramatically expanded and militarized in New York City (NYC) following the

DOI: 10.4324/9780429400346-2

attack on the World Trade Center. Stephfon narrates his encounter with the police on his own terms, raising critical questions regarding criminalisation at the heart of today's social movements, and demanding an end to the discriminatory policing of communities of colour. Taking power, he speaks back to the state that would silence him – "Shut up!" – and speaks his truth, confronting anti-Black surveillance, refusing to just be quiet. Stephfon was not acting alone when he called out the police, but as part of an intergenerational network of organisers, young people, scholars, and community residents who came together over a decade ago to "watch the watchers" and confront the racialized surveillance of the state (Browne, 2015) through the critical participatory action research (CPAR) project, Growing Up Policed.

This chapter explores critical participatory ethics as a collective project of care resonant with activist struggles against anti-Black state violence. Beyond doing no harm (one of the central precepts of ethical protocols), "a PAR-inspired understanding of social justice suggests that it is in fact unethical to look in on circumstances of pain and poverty and yet do nothing" (Manzo & Brightbill, 2007, p. 35). A critical participatory ethics extends this, insisting that first, we must attend to structures of oppression, to how the circumstances of pain and poverty are produced in the first place (see Ritterbusch, Chapter 9). A structural analysis of policing involves consideration of the historical, social, and political economic context, to understand how young people navigate racial capitalism in their everyday lives. Second, drawing upon critical theory from Black, Indigenous, scholars of colour, queer, trans and feminist theorists who attend to social science's complicity in "damage-centered research" (Tuck, 2009), a critical participatory ethics counters institutional ethics with an emphasis upon relationality and an ethics of care, centring the wellbeing of the community (Barnes, Brannelly, Ward & Ward, 2015; Battiste, 2016; Cahill, Quijada Cerecer, Reyna Rivarola, Hernández Zamudio, & Alvarez Gutiérrez, 2019; Cahill, Sultana, & Pain, 2007; DeNicolo, Yu, Crowley, & Gabel, 2017; Ellis, 2007; Fine, Weis, Weseen, & Wong, 2000; Gilligan, 1982; Guishard, Halkovic, Galletta, & Li, 2018; Hobart & Kneese, 2020; Luttrell, 2019; McKittrick, 2021; Nagar, 2019; Reyna Rivarola & López, 2021; Ritterbusch, 2012, 2019; Torre, Stoudt, Manoff, & Fine, 2017; Tuck & Guishard, 2013; Valenzuela, 1999).

Growing Up Policed began in 2012, the year following the peak of stop-and-frisk policing[1] in the context of New York City's police reform movement, at the dawn of the Black Lives Matter movement, and just before uprisings in Ferguson and elsewhere in 2014, when people across the country organised to challenge anti-Black state violence. Our research was inspired by a similar impulse – to do something! – to mobilise and decriminalise Black and Brown communities (Camp & Heatherton, 2016; Dozier, 2019; Gilmore, 2017; Kaba, 2021; Loyd, 2012; McKittrick, 2011; McKittrick & Woods, 2007; Pulido, 2017; Purnell, 2021; Sharpe, 2014; Woods, 2017). Specifically, our work contributed to organizing to end stop-and-frisk policing, and its guiding theory of public safety known as "broken windows" (Wilson & Kelling, 1982). Broken windows policing is defined by a reliance on the frequent use of surveillance practices such as stop-and-frisk and asking for identification, aggressively cracking down on minor offenses such as jumping a subway turnstile, selling loosies (single cigarettes), open alcohol containers, and biking on sidewalk. Our goal was to document the impact of broken windows policing on young people of colour, their families and communities, and to understand how young people navigate, make sense of, and resist the hypersurveillance of discriminatory policing in gentrifying New York City (Cahill et al., 2017, 2019; Stoudt et al., 2016; Stoudt, Chapter 5).

In this chapter, I trace how critical participatory ethical commitments take shape in the context of the Growing Up Policed project, challenging the assumptions underlying the theory of broken windows that has guided NYPD policing strategies for almost 40 years. Reframing how we understand community safety, Growing Up Policed shifts the problem off the backs of individuals and onto systems, structures, and policies (Torre, Fine, Stoudt, & Fox, 2012). Keeping the gaze on power, Stephfon provides an anatomy of a routine stop-and-frisk in his testimonial, calling into question the legitimacy of surveillance and data collection practices that operate as part of pervasive disciplinary regimes of knowledge and power. Not only did Stephfon have to contend with unwanted hands of the state on his body when they stopped and frisked him, but when they ran his identification, the police entered Stephfon into the "system," the New York City Police Department (NYPD) CompStat database. CompStat, a quantitative crime management, surveillance, and mapping strategy introduced by Police Commissioner Bratton in 1994, not only tracks crime but produces it, in the digitising metric-driven "big data" carceral state (Browne, 2015; Hall et al., 1978; Scannell, 2018).

Committed to expanding "the analytic frame" (McKittrick, 2021), the Growing Up Policed project offers a grounded critique of the big data capture of broken windows policing, instead centering the needs and concerns of those most affected. Youth researcher Darian X raises questions along these lines in his testimony to the City Council (2015): "Can we track human dignity? No, we cannot. But we can definitely track a community's progression and growth. Are there more institutions that support this community? . . . How are we progressing as a people – totally and holistically?" Reframing how we might understand public safety through the lens of community investment, Darian X foregrounds community needs, implicitly critiquing the narrow scope of dataveillance policing and "damage-centered" research and policies (Tuck, 2009). In the balance of the chapter, I offer an overview of critical participatory ethics in counter distinction to institutional ethics, followed by a discussion of how these commitments take shape in the Growing Up Policed research project. In conclusion, I circle back to Darian's public testimony, reflecting upon the intertwining of epistemology, theory, and ethics in our work.

Unsettling Institutional Ethics

Critical participatory praxis offers a counterpoint to debates about ethics in higher education institutions. Within the university context, ethics are governed by Institutional Review Boards (IRBs) and focus upon the potential harms and benefits of research, the rights of individual participants to information, privacy, confidentiality, and the integrity of researchers. While those considerations are not unimportant, the decontextualized narrow framing of ethics is far more attentive to issues of *accountancy* than *accountability and care*. Calls are growing to "decenter the IRB as the arbiter of ethics" (Tuck & Guishard, 2013, p. 6). Critical scholars argue that we must engage with a much more expansive framework beyond the regulatory regimes of the university, towards one that is responsive and accountable to relationships, community concerns, power, and structural inequities (Battiste, 2016; Cahill et al., 2007; Cahill & Torre, 2007; Fine & Berreras, 2004; Glass et al., 2018; Guishard et al., 2018; Nagar, 2019; Patel, 2016; Pulido, 2002; Sandwick et al., 2018; Torre et al., 2012).

There are several lines of critique calling attention to the complicity of institutional structures in reproducing harm. The first is that IRBs are interested in protecting institutions and their legal liabilities at least as much as protecting participants (Bradley,

2007; Guishard et al., 2018; Martin, 2007, Martin & Inwood, 2012; Sabati, 2019). Related to this is the concern that ethics as a regulatory regime tends to collapse ethical practice into a moment in time – the signing of an informed consent form to protect individual autonomy – rather than a relational understanding of knowledge production, accountability, ownership, vulnerability, and responsibility (Tuck & Guishard, 2013). Along these lines Fine, Weiss, Weseen, and Wong (2000) ask rhetorically, who's informed and who's consenting? And, does informed consent "mean that [respondents'] stories (and aspects of their lives they choose – or feel compelled – to share) no longer belong to them?" (p. 115). Michelle Fine reminds us, "stories of lives and relations are not sitting there like low hanging fruit, ready for the picking. You have to work with the community to determine what is sacred, what will not be documented, reported, defiled" (Fine, Torre, Burns, & Payne, 2007, p. 160).

These pointed questions about the potential of research to appropriate and erase lived experience are further unsettled when we take into consideration the broader context of structural violence and the racialised settler-colonial histories of higher education institutions (Glass et al., 2018; Kelley, 1998; McKittrick, 2014; Sabati, 2019; Tuck & Gorlewski, 2016; Tuck & Guishard, 2013; Wilder, 2013). Sheeva Sabati (2019) argues:

> Not only was the wealth of many early campuses built through practices of land and labour extraction, but colleges and universities also played an important role in consolidating ideas of racial difference as scientific truth to legitimize racialized violence within the broader social and political development of the United States.
>
> (p. 1059)

Viewed from this broader perspective, IRB review processes that define ethics as something that the researcher controls not only reproduce but also sustain settler legacies of data extraction and privatisation for profit (cf. Tuck & Guishard, 2013). In practice, the IRB functions as a gatekeeper, and this often presents difficulties for emergent collaborative projects driven by community concerns to obtain ethical clearance (Tuck & Guishard, 2013). As currently constituted, the narrow paradigm of institutional ethics authorises elite knowledge production under what Orlando Fals Borda (2001) calls "the wig of neutrality" (p. 17), that simultaneously delegitimises other forms of knowledge production. With Audre Lorde's (2003, p.111) famous words in mind, "the master's tools will never dismantle the master's house," clearly, we need new tools if we are to unsettle institutional ethics.

Challenging the erasure and extraction of knowledge, critical scholarship attends to the long history of universities using the communities of working-class Black, Indigenous, and people of colour (BIPOC) as laboratories for research, while at the same time misrepresenting and dehumanising their lives (Fine et al., 2000; Kelley, 1997, 1998; King, 2017; McKittrick, 2011; Pulido, 2002; Smith, 2013; Tuck & Guishard, 2013; Tuck & Yang, 2014). Scholars trace the inextricable connections between research, imperialism, and colonialism. As Linda Tuhiwai Smith (1999) explains, "the word itself, 'research', is probably one of the dirtiest words in the indigenous world's vocabulary" (p. 1). More recently, an important body of literature written by, and drawing upon, BIPOC feminist and critical race theorists has deepened our understandings of ethics, asking what a decolonial ethical praxis involves while calling attention to social science's complicit role in perpetuating harm (Battiste, 2016; Gilmore, 2017; Guishard & Tuck, 2014; Kaba, 2021; King, 2017; McKittrick, 2011, 2021; Nagar, 2019; Pulido, 2002; Tuck, 2009; Tuck & Guishard, 2013). Following in the footsteps of Eve Tuck (2009), who called for researchers to "suspend damage-centered research," my discussion of critical participatory ethics attends to the "recognition that some communities – particularly Indigenous, ghettoized, and Orientalized communities – are over-coded, that is, simultaneously hyper-surveilled and invisibilised/made invisible by the state, by police, and by social science research" (Tuck & Yang, 2014, p. 811).

Contesting the "apartheid of knowledge" (Bernal & Villalpando, 2002) in the academy is foundational to the explicit emphasis on racial equity and participation in critical participatory ethics. Joyce King (2017) defines the disappearance and erasure of Black knowledge as "epistemological nihilation," arguing that "one reason that state-sanctioned, racialized violence persists and with impunity is because epistemological nihilation justifies a group's physical *an*nihilation" (p. 213). In line with King's cutting analysis, the Growing Up Policed project explicitly foregrounds the insights and knowledge of young working-class people of colour's experience in NYC's public spaces. As Growing Up Policed youth researcher Markeys explains, the police make them "feel like unsafe, insecure. It makes me feel like I shouldn't even come outside anymore if I'm just gonna get harassed by a policeman that's supposed to be protecting me." Markeys says he stays home to stay safe from the police, raising questions for our research as to how broken windows policing functions as a mechanism of social and spatial control to disappear young people from neighbourhoods in gentrifying New York City (Cahill et al., 2017; Cahill et al., 2019).

Challenging the "colonial unknowing" that obscures the academy's racial entanglements and settler-colonialism (Sabati, 2019; Tuck & Yang, 2014), a critical participatory ethics attends to historic harms perpetuated by social research. In our work, this meant that we analysed how "damage-centred" (Tuck, 2009) culture of poverty theories (i.e., "tangle of pathology"), underlie not only broken windows policing, but urban development policies of "benign neglect" and state abandonment (Kelley, 1997; Moynihan, 1965; Wallace & Wallace, 1998; McKittrick, 2011; Gilmore & Gilmore, 2016). Collectively, we traced the impact of catastrophic disinvestment in redlined[2] working class communities of colour across NYC and the country (Wallace & Wallace, 1998; Fullilove & Wallace, 2011; Coates, 2015). And at the same time, we attend to community networks of kinship and care that are crucial to navigating and resisting state and slow violence, as we shall discuss (Gilmore, 2017, 2022; Kaba, 2021; Purnell, 2021; Spade, 2020).

Critical Participatory Ethics

How might critical participatory ethics provide new insights for reforming ethical review board structures so they encourage, rather than restrict, liberatory collaborative research projects? Not a method, critical PAR is an epistemological stance that has profound implications for rethinking ethical commitments, which in turn raise new questions for theory and practice. Rooted in principles of justice and radical democracy, PAR is an inclusive, collaborative approach to knowledge production. PAR is defined by commitments to deep participation to produce knowledge in the interest of social change (Fine & Torre, 2021; Torre, Cahill, & Fox, 2015). Placing emphasis on the *critical*, critical participatory action research (CPAR) signals explicit attention to the "uneven structural distribution of opportunities, resources, and dignity; troubles ideological categories projected onto communities (delinquent, at risk, damaged, innocent, victim); and contests how 'science' has been recruited to legitimate dominant policies and practices" (Torre et al., 2012, p. 171). CPAR opens up a critical space for challenging hegemonic discourses, reframing how we understand the problem informed by a range of critical social theories concerned with structural inequities and power, including queer, Marxist, critical race, feminist, Indigenous, and decolonial theories, amongst others (Torre et al., 2012).

At its most foundational, critical participatory ethics is guided by people's right to research. Appardurai (2006) argues that within the political economic context of globalisation, which for many is a state of ongoing crisis, research is a right to "systematically increase the stock of knowledge which they consider most vital to their survival as human beings and to their claims as citizens" (p. 168). This right is the focus of our work. Not only do young working-class people of colour, who are most impacted by discriminatory policing, hold deep knowledge, but their particular vantage point from the front line is fundamental to understanding structural inequalities. Informed by an "intimate geopolitics" (Pain & Staeheli, 2014), critical participatory ethics emphasises the relationships between social structures and injustice in everyday life experiences, engaging multiple methods that triangulate the personal and the political (Cahill, 2007).

Critical participatory ethics unfold relationally rather than top-down, explicitly negotiating asymmetries of power, privilege, and knowledge production as part of the research process, excavating fractures and dissent (Askins & Pain, 2011; Nagar, 2014; Sultana, 2007; Torre, 2006). As Guishard et al. (2018) argue in their call for an ethics of intersubjectivity, "members of communities hold knowledge of their subjective experience, and live and navigate systems that are entrenched in hierarchies of power, noting that we as researchers are also entangled in these hierarchies of power" (p. 19). Participatory ethics are understood to be emergent, culturally, and contextually specific, situated, and accountable to place and community (in all of its heterogeneity). Nagar (2019) captures this in her:

> insistence on a collective ethic of radical vulnerability . . . that commits itself to grappling with an always fluid and unresolvable set of incommensurabilities in ongoing relationships where the meanings of 'justice', 'ethics', or 'politics' can only emerge in the shifting specifics of a given moment in an ongoing struggle.
>
> (p. 18)

While many researchers choose to engage in PAR precisely for ethical reasons, Manzo and Brightbill (2007) caution that doing so does not circumvent ethical dilemmas, and indeed it may raise new dilemmas. The scholarship focused on participatory ethics offers richly detailed accounts of ethical quandaries, critically interrogating the tensions, providing meaningful insights for practice and theory as well as examples of developing ethical frameworks as a collective (Baloy, Sabati, & Glass, 2016; Banks et al., 2013; Billies, Francisco, Krueger, & Linville, 2010; Cahill et al., 2007; Elwood & Leszczynski, 2011; Glass et al., 2018; Guishard, 2015; Guishard et al., 2018; Krueger, 2011; Manzo & Brightbill, 2007; Nagar, 2019; Pain et al., 2015; Ritterbusch, 2012, 2019; Tuck & Guishard, 2013). Consideration is given to relationships, representation, power dynamics, and structural inequities. Scholars flag significant concerns about romanticising and tokenizing the community, raising questions about who speaks for and represents community concerns (Cahill & Torre, 2007; Reyna Rivarola & López, 2021; Spivak, 2007). Tuck and Yang's (2014) critical theorisation of "refusal" is particularly relevant to critical participatory ethics as an analytic practice that centres power:

> Refusal, and stances of refusal in research, are attempts to place limits on conquest and the colonization of knowledge by marking what is off limits, what is not up for grabs or discussion, what is sacred, and what can't be known.
>
> (p. 225)

Refusal spotlights the structural harms of social research, moving beyond an individual not signing a consent form (Tuck & Yang, 2014). Drawing on the wisdom of anti-apartheid and disabilities movements, stating that "nothing about us without us is for us," a critical participatory ethics takes inspiration from collective efforts to reform ethical review processes. An illuminating example is the Bronx Community Research Review Board (http://bxcrrb.org), which centres the voices of residents, ensures appropriate representation, and protects Bronx communities from "academic research abuse" (Guishard, 2015).

In what follows, I discuss the Growing Up Policed project, drawing out critical participatory ethical commitments as they took shape in our work. I highlight our explicit attention to structural analysis grounded in lived experience and collective inquiry, an ethics of care that foregrounds relationships as part of our process, as well as the politics of representation.

Growing Up Policed as a Project of Care

> I was walking. I had a hoodie on, and the cops just started looking at me weird. He came up to me, went in my pockets. And like, in my pocket, in my left pocket, he put his hand in there and then he started patting me down, putting the flashlight around me, started searching me with the flashlight to see if he could find anything, and didn't. So they left. The cops said I looked suspicious. I was, like, thirteen years old and I felt uncomfortable, it was, like, a bad experience. When the officers searched me that time, I wish they could've talked to me first about how I looked suspicious, instead of just searching me. He could've just told me the reason why I look suspicious, and asked me if I had anything on me, things like that. . . .
>
> (Jesus, age 15, Brooklyn)

The 1994 missive *Police Strategy No. 5: Reclaiming the Public Spaces of New York City*, co-authored by Mayor Giuliani and Police Commissioner Bratton, identified teenagers hanging out on the street as a sign of disorder to be ordered. This heralded the notorious quality-of-life campaign that made broken windows policing famous around the world. While these teenagers' race and gender were not made explicit in the report, "urban youth" was coded as Black, Brown and male in the "inner city." Criminalised representations of young people of colour as "visible signs of a city out of control" (Giuliani & Bratton, 1994, p. 5) were, and continue to be, central to producing and securitising the "crisis at home" (Katz, 2017), in the gentrifying but still disinvested neighbourhoods of New York City (Cahill et al., 2019). Broken windows policing promoted a "return to urban civility" in the context of the 1980s "urban crisis," a crisis manufactured by decades of state policies of segregation and abandonment (Berman & Berger, 2007; Cowen & Lewis, 2016; McKittrick, 2011; Wallace & Wallace, 1998). Broken windows Focused on "order-maintenance" (Wilson & Kelling, 1982), broken windows policing was initially credited with producing a precipitous drop in crime in NYC in the 1990s. However, critical criminology scholars have debunked this, demonstrating how broken windows policing is not evidence-based, but instead produces a way of "seeing disorder" that further reproduces urban racial inequalities (Harcourt, 2009; Sampson & Raudenbush, 2004; Smith, 2001). A federal class action lawsuit came to the same conclusion, ruling that the NYPD's stop-and-frisk practices were a form of racial profiling and unconstitutional (Floyd vs. City of New York, 2013).

Narrating his bad experience with the police, Jesus shares his discomfort and sense of violation as the cop goes through his pockets without his consent. Following up with "I wish . . . ," Jesus articulates what could and should have been, centring desire and opening up an alternative imagination for what a more respectful encounter with the police might be like. This sense of desire animates our participatory research. On the frontlines of decriminalising their own lives, young people are leading social movements and challenging racialised surveillance. And yet too often young people's perspectives are conspicuously missing from policy decisions that significantly impact their lives, as well as much of the scholarship focused on policing (Beckett & Herbert, 2010; Billies, 2016; Bustamante, Jashnani, & Stoudt, 2019; Camp & Heatherton, 2016; Dozier, 2019; Fine et al., 2003; Gilmore, 2007; Kaba, 2021; Loyd & Bonds, 2018; Purnell, 2021; Ritchie, 2017; Stoudt, Fine, & Fox, 2011; Stoudt et al., 2019; Vitale, 2017). Our research sought to rectify this erasure, foregrounding young people's critical insights through collective knowledge production.

The Growing Up Policed project developed out of a series of discussions between organisers and scholars to consider how research might be mobilised as part of the citywide movement to end stop-and-frisk policing (Stoudt et al., 2016). The intergenerational PAR project is a partnership between the Youth Power Project of Make the Road New York (see https://maketheroadny.org/) and the Public Science Project at the Graduate Center, City University of New York. Significantly, organisers from the Youth Power Project have been actively addressing discriminatory policing for many years in advance of our collaboration, including doing extensive outreach, education, and organising with young people about their rights in schools and communities. Their intimate knowledge and experiences were foundational in framing the Growing Up Policed project. The critical participatory ethics that informed our project centred questions of accountability and social justice as an explicit part of our process. This is what Torre et al. (2012) describe as "impact validity" (p. 180), anticipating from the beginning how to establish a network of allies and how to produce evidence that can be mobilised for change and in solidarity with social movements, reaching out to diverse audiences/publics "beyond the journal article" (Cahill & Torre, 2007).

Starting with the concerns and questions of youth organisers, our project engaged collectively in an ongoing process of dialogue and critical reflection informed by ethical and epistemological commitments to inclusion (Freire, 1997). Our research design encompassed a range of both qualitative and quantitative methodologies to triangulate structural conditions and history with first-hand intimate experiences with the police. Methods included reanalysing and counter-mapping public NYPD data, as well as surveys, video testimonials, and focus groups. We also used a range of creative methods, including a participatory documentary (*Who's Impacted by Stop & Frisk?*), comic strips, spoken word performances, and an exhibition (Stoudt et al., 2016). The research process itself created opportunities for engagement and education. For example, our exhibition "More than a Quota" provided an opportunity to get feedback from young people and engage with them in dialogue about the research findings (see www.growinguppoliced.org).

A critical participatory ethics centres a relational praxis. Through collaboration, we build relationships with each other and do so in community. As the young people from Make the Road New York were first and foremost organisers, we considered throughout the process how our research is accountable to, and in dialogue with, activist movements, and in turn how activism informs our analysis. Our theories of change shifted over the course of our work together. Originally, our project was titled the Researchers for Fair Policing, motivated by a desire for justice, and what with hindsight might be characterised as a sense of "cruel optimism" (Berlant, 2011). At first, "fair policing" was not understood to be a contradiction by the young people involved in the project, which began at the dawn of the Black Lives Matter movement. But over time, as we analysed the damning data of stop-and-frisk, the patent unfairness and injustice of policing was thrown into sharp relief, making clear the urgency of attending to a structural analysis as part of our critical participatory ethics. Emotion – grief, outrage – and collective engagement in public protest movements following the police killings of Mike Brown, Eric Garner, Tamir Rice, Akai Gurley, Rekia Boyd, and far too many others, shifted our collective consciousness. This shift moved us towards an abolitionist framework informed by an analysis of how racial capitalism takes shape in gentrifying/still disinvested Brooklyn (Cahill et al., 2019; Gilmore, 2022; Wilder, 2001).

We Watch the Watchers

At the beginning of our project, we focused on quantitative data, as the initial purpose of our project was to challenge broken windows policing. How might we respond to the big data-driven enforcement of broken windows policing/Compstat with our own data and re-represent the problem? Reanalysing publicly available NYPD data, Brett Stoudt and co-researchers created "the geography of stop & frisk," an animated counter-map clearly demonstrating the disproportionate targeting of Black and Brown neighbourhoods (see www.GrowingUpPoliced.org). Refusing to circulate hypervisible images of police violence that dominate the media, we instead shifted the gaze, making "transparent the metanarrative of knowledge production – its spectatorship for pain and its preoccupation for documenting and ruling over racial difference" (Tuck & Yang, 2014, p. 817). Unhiding how stop-and-frisk functions as a form of surveillance and data collection, the counter-mapping reveals how policing works on the ground to produce the spectacle of criminalisation, that in turn reinforces the justification for policing (see Cahill et al., 2019 for further discussion). The number of recorded stop-and-frisks increased over 600% during the tenure of Mayor Bloomberg (2002–2013), during which time over a third of the city was rezoned, "upzoning" communities of colour for development (Angotti & Morse, & Stein, 2016; Stein, 2019). Mapping the police data brought us into conversation with the slow violence of racial capitalism. We noticed how the very neighbourhoods that were overpoliced were the same neighbourhoods that were disinvested, and now gentrifying (Cahill et al., 2019). Lisa Maria Cacho (2012) writes that for those who are criminalised, they are "not just excluded from justice . . . they and the places where they live . . . are imagined as the reason why a punitive (in)justice system exists" (p. 5; cf. Hall et al., 1978). Counter-mapping the NYPD data in conjunction with a historical analysis and the intimate first-hand experiences of police encounters demonstrated how property regimes produce and securitize racialized space (Bonds, 2019; Cahill et al., 2019; Camp, 2009; Correia, 2013; Gilmore, 2007; McKittrick, 2011; McKittrick & Woods, 2007; Pulido, 2016; Roy, 2017).

We collected over 1,000 surveys from young Black and Latinx people across the city that document the oppressive presence of police in the most personal spaces of young people's everyday lives: in their neighbourhood streets, schools, and homes. A total of

89% of the young people surveyed had some *personal* contact with police and 87% witnessed the police stop their family, friends, or neighbours.

Young people of colour reported feeling that they are constantly under surveillance and being watched all the time by the cops. They are asked for IDs in the lobby of their buildings, told to move when standing on a street corner, and stopped and patted down on their way to the corner store. Cops patrol their school hallways. One of the things we learned through our research is that many young people did not know their rights. So, at the same time that youth researchers/organisers distributed the survey, they also conducted "Know Your Rights" trainings, in keeping with our ethics of care and commitment to building capacity (Stoudt et al., 2016).

The video testimonials, like the counter-mapping of stop-and-frisk, offer another public platform for "watching the watchers" (Browne, 2015). Collecting young people's stories of encounters with the police, we created a "living" archive, informed by the feminist Chicanx testimonios tradition that engages in story-sharing as an approach to building solidarity with an emphasis on embodied knowledge (Delgado Bernal, Burciaga, & Flores Carmona, 2012; Moraga & Anzaldúa, 1983). The testimonials reflect what Simone Browne (2015) articulates as "dark sousveillance,"[3] as an "imaginative place from which to mobilize a critique of racializing surveillance . . . and other freedom practices . . . plotting imaginaries that are oppositional and that are hopeful for another way of being" (p. 21). Challenging the criminalisation of young people of colour, a critical participatory ethics explicitly engages with the politics of representation, contesting what Foucault (1980) identified as the "subjectifying social sciences." Jesus, Stephfon, and many others took great risks to share their stories in their testimonials, "calling out" the police, while "calling in" a broader public who do not usually get to hear young people share their experiences of their encounters with the police on their own terms. Viewing their stories together, we bear witness to what McKittrick (2011) characterizes as the "relational and connective life-force . . . [of] 'ordinary people' who collectively live crisis" and how "human life is analytically understood as both ordinary and underwritten by mortal urgency" (p. 959). Collectively the video testimonials reflect our participatory process of moving from personal experiences to social theorising to develop a shared critical analysis of unjust conditions, in this case understanding discriminatory policing as systemic and responding by taking action, as an ethical imperative.

Documenting Care

> On my block there are these groups of boys . . . when I see them, when I see the police, I have my phone ready to record. I'm just waiting for something to happen. And it's, like, not even just them as, like, that's like my older brother too. . . . It's been known that he can't leave the corner without being stopped by police. So that's one of the things. He's not safe neither. . . . I'm scared that his life will be taken away.
>
> (Jasmine, 17, Brooklyn)

At the ready, armed with her phone, Jasmine is "engaging in care as shared risk" (Sharpe, 2016, p. 180). She is "watching the watchers" too (Browne, 2015). While young men of colour are disproportionately targeted by the police, one of the key findings of our work is that broken windows policing is not experienced individually, but collectively as an assault on the community. While broken windows policing was packaged as quality-of-life reforms, not surprisingly, our research demonstrates the opposite. Broken windows policing produces disorder in the lives of young people and their families, wearing out the fabric of community relationships, and curtailing the freedom of movement in one's own neighbourhood by creating an overall hostile environment. This sense of disordering is compounded by a convergence of structural dispossessions. While young people are being stopped by the police, rents are going up, families are being displaced, good jobs that pay a living wage are scarce, and funding is cut from schools and public housing (Cahill et al., 2017; Stoudt et al., 2016).

In this context of "organised abandonment" and the "organised violence that it depends upon" (Gilmore & Gilmore, 2016, p. 198), our work documents how people come together to meet each other's survival needs in networks of care and mutual aid (Kaba, 2021; Spade, 2020). The Growing Up Policed project may be understood along these lines as an intergenerational collective coming together to resist and mobilize against the police state. A critical participatory ethics centres relationships that are founded on what DeNicolo et al. (2017) describe as "*cariño conscientizado*," a critically conscious care praxis that seeks to dismantle structural injustice (Bartolomé, 2008; Cahill, Quijada Cerecer et al., 2019; Freire, 1997; Rolón-Dow, 2005; Valenzuela, 1999). Following Ruth Wilson Gilmore, who reminds us that "abolition is about presence, not absence. It's about building life-affirming institutions" (Kumanyika, 2020). Our analysis attends to how people create community and support each other in the context of the slow violence of state abandonment and neoliberal austerity, in neighbourhoods across Brooklyn, where Stephfon, Jesus, and Jasmine live with their families, navigating state surveillance. This includes local organisations, such as our community partner Make the Road New York, which developed to support immigrant communities' wellbeing, community development, and legal services in a context of targeted disinvestment and catastrophic state abandonment in Bushwick in the 1990s (Cahill et al., 2019; Fullilove & Wallace, 2011; Gilmore & Gilmore, 2016; Wallace & Wallace, 1998). The Youth Power Project formed to build capacity to organise and act on the urgent issues facing young people and their communities. As a project of care, our research is inspired by those who, over many years, continue to organise tirelessly behind the scenes, in community and in protest, informed by an ethics of *cariño conscientizado* (DeNicolo et al., 2017).

Also of significance are the informal ways that extended family, friends, and community members show up, like Jasmine, camera in hand. Jasmine knows her brother's friends and they know her. They are all part of the community that is looking out for each other, like the abolitionist refrain, "We Keep Us Safe." "Eyes on the street" keep the community safe (Cahill et al., 2017; Jacobs, 1961): the mothers who watch out windows or monitor building lobbies (Stoudt et al., 2019; Torre et al., 2017), the teachers who wait outside of schools to make sure their students are not detained by the police, and the neighbours who advocate for each other and watch over the kids on the block. Communities that have collectively endured disinvestment and hostile policing look out for and take care of each other in long-standing traditions of mutual aid to address survival needs (Billies, 2015; Gilmore & Gilmore, 2016; Spade, 2020; Welfare Warriors Research Collaborative, 2010). As part of creating the conditions for the "real estate state" (Stein, 2019), the police and gentrification collude to push Black and Brown young people out of public spaces (Cahill et al., 2019). Our research raises concerns about how community care and support networks are threatened in the process. As more families are displaced in gentrifying Brooklyn, there are fewer eyes on the street watching out for each other. One of our key recommendations along these lines is to shift funding from policing budgets and invest in communities.

Conclusion

> If – supposition, hypothesis, possibility, requirement,
> stipulation.
> – a location to live in if, to live as if
> – To live (as if) in a future time
> – as if . . . Black living was not interdicted,
> as if the streets were not militarized and organised
> against Black life, Black gathering, Black being, Black
> breath, Black habitation.
> To live as if there is enough.
>
> Christina Sharpe (2021, p. 25)

At the heart of a critical participatory ethics is a collective process for reckoning with structural inequities while holding a vision for justice. A vision for what could and should be: "to live as if there is enough" (Sharpe, 2021, p. 25). Katherine McKittrick (2014) argues we must "foster a commitment to acknowledging violence and undoing its persistent frame" (p. 18). Along these lines, a critical participatory ethics asks: How do we represent violence in a way that does not reinscribe harm? Throughout our participatory process we revisited the question of how to "undo the persistent frame" of violence in our research, our analysis, our writing, and in our actions. We grappled with undoing the frame in our discussions about how to speak back to damage-centred research and policies, and the criminalisation of communities of colour (McKittrick, 2011; Tuck, 2009). Collectively we considered the thorny issues of representation and action, centring the purposes and publics of our research, while revising our theories of change. What's involved with dismantling the masters house (Lorde, 2003) of broken windows policing? How might our research provoke meaningful action? *Undoing* meant explicitly raising questions about what public safety beyond policing involves. While there is now a robust public discourse about abolition, this was a new idea for many of us when we started the Growing Up Policed project in 2012, including the organisers, young people, and their families. At the time we did not have the language to frame our research along the lines of abolition or defunding the police. Instead, in our initial findings we called for rethinking police reform as community investment, and in particular investing in community-based organisations that have been on the ground addressing survival needs for decades (such as Make the Road New York).

Asking what "*if* one was attuned to sustaining, really sustaining, Black life?," Christina Sharpe (2021, p. 26) raises questions informed by the Black radical imagination, what Robin D. G. Kelley (2002) calls "freedom dreams," arguing that we must take into account not only what we are fighting against, but what we are fighting for. Circling back to Darian X's forceful testimony to the New York City Council in 2015, he draws upon our research and contests the proposal to hire 1,000 new "community" police officers. Reframing the narrative of community safety and crime, Darian X centres the everyday lived experience of young Black and Brown people within the context of disinvestment and the criminalisation of communities of colour:

> It is about total reinvestment in our communities of colour. . . . Crime is caused by lack of housing, lack of job opportunities, lack of sound educational systems and

> institutions, lack of extracurricular things for our young people to get involved in after school. After school young people are going back to homes where their parents have to work until 9, 10 o'clock at night to support them, so there is no one home when they get back there. There is nothing but the community that has been criminalised, that has been broken down systemically, so what we need to do is really uphold, support and reinvest in young people in communities of color.

Flipping the script on the state, Darian highlights their responsibilities toward young people. He continues:

> I feel like graduation rates are a great way to track outcomes. Are more of our young people graduating? Are less of our young people in prison? . . . Are the suspension rates going down? These are trackable ways of seeing if improvement is really happening in our community. Do more people have access to jobs then they did when we started this program? . . . Right now we can see that people are underemployed, undereducated, overworked, under accredited with humanity and dignity as a person. So, can we track human dignity? No, we cannot. But we can definitely track a community's progression and growth. Are there more institutions that support this community? Are there more banks that give loans to developers that build low-income houses in this community? How are we progressing as a people – totally and holistically?

Mobilizing the normative logic of "indicators," Darian proposes an alternative set of measures that centre community wellbeing (including graduation rates, jobs, and affordable housing) and challenge the carceral logic of "community policing." In so doing, he exposes the biases of NYPD's seemingly objective big data on crime that underpin broken windows policing and transform gentrifying communities into "prisonized landscapes" for Black and Brown communities (Shabazz, 2015). Questioning whether we can track human dignity and community progression, Darian gestures to the complicity of "damage-centred" social science (Tuck, 2009) in legitimating harm to communities. At the same time, he rescales the problem from pathologizing individuals to the systems, structures, and policies that not only produce racial inequality and dispossession, but criminalisation.

How might research not only be useful for communities, but co-created *with* (and not on, or for) communities, to address their urgent concerns and support their progress "totally and holistically?" Raising significant questions about the purposes and publics of research, critical participatory ethics radically shifts the ground for justice-oriented research and its implications.

Acknowledgments

I am grateful for the care, generous editorial feedback, and critical insights of Mike Kesby in particular – and Rachel Pain and Sara Kindon – the wonder team of participatory geographies. All scholarship is collaborative whether this is acknowledged or not. Immense appreciation to Brett Stoudt and María Elena Torre for their beautiful collaboration and conversation, for challenging, inspiring, and supporting. It has been an honour to work closely with the Growing Up Policed research team for over many years. I am grateful to Darian X, Amanda Matles, Jose Lopez, Adilka Pimentel, Selma Djokovic, and Kimberly Belmonte. We are thankful to all the young people who bravely shared their stories and engaged with others in collective analysis and action to address state violence and imagine a more just future. Special thanks to Leticia Alvarez Gutiérrez and Bevin Cahill.

This research would not be possible without the support of the Adco Foundation; Antipode Foundation; Graduate Center, CUNY; Institute for Human Geography; Taconic Fellowship, Pratt Center for Community Development; and the Sociological Initiatives Foundation.

Notes

1 Stop-and-frisk is a policing strategy where "police officers – providing they could 'point to specific and articulable facts' . . . – were allowed to temporarily detain a citizen's freedom to ask questions and do so without a warrant or 'probable cause' but instead the lesser 'reasonable suspicion' that a crime has, is, or about to be committed" (Stoudt, Fine & Fox, 2011, p. 1333). In 2013, District Court Judge Shira A. Scheindlin ruled that the way stop-and-frisk was carried out by the NYC Police Department was unconstitutional and racial profiling.

2 Redlining is the illegal practice of banks and other institutions refusing to offer credit, mortgages, or insurance in particular neighborhoods based on racial and ethnic composition. Although redlining was outlawed by the 1968 Fair Housing Act, it continues in various forms today. It is an enduring and clear example of structural racism in US history (Rothstein, 2017).

3 "Sousveillance" is an act of observing and recording, often using personal devices, actively inverting the power relations of surveillance (Mann, Nolan, & Wellman, 2003).

References

Angotti, T., Morse, S., & Stein, S. (2016). *Zoned Out! Race, Displacement, and City Planning in New York City*. New York: Terreform.

Appadurai, A. (2006). The right to research. *Globalisation, Societies and Education*, 4(2): 167–177.

Askins, K., & Pain, R. (2011). Contact zones: Participation, materiality, and the messiness of interaction. *Environment and Planning D: Society and Space*, 29(5): 803–821. https://doi.org/10.1068/d11109

Baloy, N. J. K., Sabati, S., & Glass, R. D. (2016). *Unsettling Research Ethics: A Collaborative Conference Report*. Santa Cruz: UC Centre for Collaborative Research for an Equitable California.

Banks, S., Armstrong, A., Carter, K., Graham, H., Hayward, P., Henry, A., Holland, T., Holmes, C., Lee, A., McNulty, A., & others. (2013). Everyday ethics in community-based participatory research. *Contemporary Social Science*, 8(3): 263–277.

Barnes, M., Brannelly, T., Ward, L., & Ward, N. (Eds.) (2015). *Ethics of Care: Critical Advances in International Perspective*. Bristol: Policy Press.

Bartolomé, L. I. (2008). Authentic cariño and respect in minority education: The political and ideological dimensions of love. *The International Journal of Critical Pedagogy*, 1(1): 1–17.

Battiste, M. (2016). Research Ethics for Protecting Indigenous Knowledge and Heritage: Institutional and Researcher Responsibilities. In N. K. Denzin & M. D. Giardina (Eds.), *Ethical Futures in Qualitative Research: Decolonizing the Politics of Knowledge*. Abingdon: Routledge.

Beckett, K., & Herbert, S. (2010). *Banished*. New York: Oxford University Press. http://www.sjsu.edu/justicestudies/programs-events/ann-lucas/beckett-k/Banished-Lecture.pdf

Berlant, L. G. (2011). *Cruel Optimism*. Durham, NC: Duke University Press.

Berman, M., & Berger, B. (Eds.) (2007). *New York Calling: From Blackout to Bloomberg*. London: Reaktion Books.

Bernal, D. D., & Villalpando, O. (2002). An apartheid of knowledge in academia: The struggle over the "legitimate" knowledge of faculty of color. *Equity & Excellence in Education*, 35(2): 169–180.

Billies, M. (2015). *Low Income LGBTGNC (Gender Nonconforming) Struggles over Shelters as Public Space.* New York: City University of New York. https://academicworks.cuny.edu/cgi/viewcontent.cgi?article=1159&context=kb_pubs

Billies, M. (2016, 1 March). Impossible compliance: Policing as violent struggle over bodies and urban space. *Metropolitics.* http://www.metropolitiques.eu/Impossible-Compliance-Policing-as.html

Billies, M., Francisco, V., Krueger, P., & Linville, D. (2010). Participatory action research: Our methodological roots. *International Review of Qualitative Research*, 3(3): 277–286.

Bonds, A. (2019). Race and ethnicity I: Property, race, and the carceral state. *Progress in Human Geography*, 43(3): 574–583.

Bradley, M. (2007). Silenced for their own protection: How the IRB marginalizes those it feigns to protect. *ACME: An International Journal for Critical Geographies*, 6(3): 339–349.

Browne, S. (2015). *Dark Matters: On the Surveillance of Blackness.* Durham, NC: Duke University Press.

Bustamante, P., Jashnani, G., & Stoudt, B. G. (2019). Theorizing cumulative dehumanization: An embodied praxis of "becoming" and resisting state-sanctioned violence. *Social and Personality Psychology Compass*, 13(1): 1–13. https://compass.onlinelibrary.wiley.com/doi/10.1111/spc3.12429

Cacho, L. M. (2012). *Social Death: Racialized Rightlessness and the Criminalization of the Unprotected.* New York: NYU Press.

Cahill, C. (2007). The personal is political: Developing new subjectivities through participatory action research. *Gender, Place and Culture*, 14(3): 267–292.

Cahill, C., Quijada Cerecer, D. A., Hernández Zamudio, J., Reyna Rivarola, A., & Alvarez Gutiérrez, L. (2019). 'Caution, we have power': Resisting the school-to-sweatshop pipeline through participatory artistic praxes and critical care. *Gender & Education*, 31(5): 576–589.

Cahill, C., Stoudt, B. G., Matles, A., Belmonte, K., Djokovic, S., Lopez, J., Pimentel, A., Torre, M. E., & Darian, X. (2017). The right to the sidewalk: The struggle over broken windows policing, young people, and NYC streets. In J. Hou & S. Kneirbein (Eds), *City Unsilenced: Urban Resistance and Public Space in the Age of Shrinking Democracy.* 94–105. New York: Routledge,

Cahill, C., Stoudt, B. G., Torre, M. E., Darian, X., Matles, A., Belmonte, K., Djokovic, S., Lopez, J., & Pimentel, A. (2019). "They were looking at us like we were bad people": Growing up policed in the gentrifying, still disinvested city. *ACME: An International Journal for Critical Geographies*, 18(5): 1128–1149.

Cahill, C., Sultana, F., & Pain, R. (2007). Participatory ethics: Politics, practices, institutions. *ACME: An International E-Journal for Critical Geographies*, 6(3): 304–318.

Cahill, C., & Torre, M. E. (2007). Beyond the journal article: Representations, audience, and the presentation of participatory action research. In S. Kindon, R. Pain, & M. Kesby (Eds.), *Connecting People, Participation and Place: Participatory Action Research Approaches and Methods.* 196–206. London: Routledge.

Camp, J. T. (2009). "We know this place": Neoliberal racial regimes and the Katrina circumstance. *American Quarterly*, 61(3): 693–717.

Camp, J. T., & Heatherton, C. (2016). *Policing the Planet: Why the Policing Crisis Led to Black Lives Matter.* New York: Verso Books.

Coates, T. (2015, 24 September). Moynihan, mass incarceration, and responsibility. *The Atlantic.* https://www.theatlantic.com/politics/archive/2015/09/moynihan-mass-incarceration-and-responsibility/407131/

Correia, D. (2013). *Properties of Violence: Law and Land Grant Struggle in Northern New Mexico.* Athens, GA: University of Georgia Press.

Cowen, D., & Lewis, N. (2016, 2 August). Anti-blackness and urban geopolitical economy reflections on Ferguson and the suburbanization of the internal colony. *Society & Space.* https://www.societyandspace.org/articles/anti-blackness-and-urban-geopolitical-economy

Delgado Bernal, D., Burciaga, R., & Flores Carmona, J. (2012). Chicana/Latina testimonios: Mapping the methodological, pedagogical, and political. *Equity & Excellence in Education*, 45(3): 363–372.

DeNicolo, C. P., Yu, M., Crowley, C. B., & Gabel, S. L. (2017). Reimagining critical care and problematizing sense of school belonging as a response to inequality for immigrants and children of immigrants. *Review of Research in Education*, 41(1): 500–530.

Dozier, D. (2019). Contested development: Homeless property, police reform, and resistance in Skid Row, LA. *International Journal of Urban and Regional Research*, 43(1): 179–194.

Ellis, C. (2007). Telling secrets, revealing lives: Relational ethics in research with intimate others. *Qualitative Inquiry*, 13: 3–29.

Elwood, S., & Leszczynski, A. (2011). Privacy, reconsidered: New representations, data practices, and the geoweb. *Geoforum*, 42(1): 6–15.

Fals Borda, O. (2001). Participatory (action) research in social theory: Origins and challenges. In P. Reason & H. Bradbury (Eds.), *Handbook of Action Research*. 27–37. London: Sage Publications.

Fine, M., & Barreras, R. (2004). To be of use. *Analyses of Social Issues and Public Policy*, 1: 175–182. http://www.asap-spssi.org/issue2.htm

Fine, M., Freudenberg, N., Payne, Y., Perkins, T., Smith, K., & Wanzer, K. (2003). "Anything can happen with police around": Urban youth evaluate strategies of surveillance in public places. *Journal of Social Issues*, 59(1): 141–158.

Fine, M., & Torre, M. E. (2021). *Essentials of Critical Participatory Action Research*. Washington, DC: American Psychological Association.

Fine, M., Torre, M. E., Burns, A., & Payne, Y. A. (2007). Youth research/participatory methods for reform. In D. Thiessen & A. Cook-Sather (Eds.), *International Handbook of Student Experience in Elementary and Secondary School*. 805–828. Dordrecht: Springer.

Fine, M., Weis, L., Weseen, S., & Wong, L. (2000). For whom?: Qualitative research, representations, and social responsibilities. In N. K. Denzin & Y. S. Lincoln (Eds.), *Handbook of Qualitative Research*. 107–131. Thousand Oaks, CA: Sage Publications.

Foucault, M. (1980). *Power/Knowledge: Selected Interviews & Other Writings 1972–1977*. New York: Pantheon Books.

Freire, P. (1997). *Pedagogy of the Oppressed*. Penguin Books.

Fullilove, M. T., & Wallace, R. (2011). Serial forced displacement in American cities, 1916–2010. *Journal of Urban Health*, 88(3): 381–389.

Gilligan, C. (1982). *In a Different Voice: Psychological Theory and Women's Development*. Cambridge, MA: Harvard University Press.

Gilmore, R. W. (2007). *Golden Gulag: Prisons, Surplus, Crisis, and Opposition in Globalizing California*. Berkeley, CA: University of California Press.

Gilmore, R. W. (2017). Abolition geography and the problem of innocence. In G. T. Johnson & A. Lubin (Eds.), *Futures of Black Radicalism*. London: Verso Books.

Gilmore, R. W. (2022). *Abolition Geography: Essays towards Liberation*. London: Verso Books.

Gilmore, R. W., & Gilmore, C. (2016). Beyond Bratton. In J. T. Camp & C. Heatherton (Eds.), *Policing the Planet: Why the Policing Crisis Led to Black Lives Matter*. 173–199. London: Verso Books.

Giuliani, R. W., & Bratton, W. J. (1994). *Police Strategy no. 5: Reclaiming the Public Spaces of New York*. https://www.ojp.gov/

Glass, R. D., Morton, J. M., King, J. E., Krueger-Henney, P., Moses, M. S., Sabati, S., & Richardson, T. (2018). The ethical stakes of collaborative community-based social science research. *Urban Education*, 53(4): 503–531.

Guishard, M. A. (2015). *Nepantla and Ubuntu Ethics Para Nosotros: Beyond Scrupulous Adherence toward Threshold Perspectives of Participatory/Collaborative Research Ethics*. Doctoral dissertation. New York: City University of New York. https://academicworks.cuny.edu/gc_etds/957/

Guishard, M. A., Halkovic, A., Galletta, A., & Li, P. (2018). Toward epistemological ethics: Centering communities and social justice in qualitative research. *Forum Qualitative Sozialforschung/ Forum: Qualitative Social Research*, 19(3) Art.9. https://doi.org/10.17169/fqs-19.3.3145

Guishard, M., & Tuck, E. (2014). Youth resistance research methods and ethical challenges. In E. Tuck & K. W. Yang. *Youth Resistance Research and Theories of Change*. 193–206. New York: Routledge.

Hall, S., Critcher, C., Jefferson, T., Clarke, J., & Roberts, B. (1978). *Policing the Crisis – Mugging, the State and Law and Order*. New York: Holmes and Meier Publishers. http://www.ncjrs.gov/ App/abstractdb/AbstractDBDetails.aspx?id=51970

Harcourt, B. E. (2001). *Illusion of Order: The False Promise of Broken Windows Policing*. Cambridge, MA: Harvard University Press.

Hobart, H. J. K., & Kneese, T. (2020). Radical Care: Survival Strategies for Uncertain Times. *Social Text*, 38(1): 1–16. https://doi.org/10.1215/01642472-7971067

Jacobs, J. (1961). *The Death and Life of Great American Cities*. New York: Vintage.

Kaba, M. (2021). *We Do This 'Til We Free Us: Abolitionist Organizing and Transforming Justice*. Chicago: Haymarket Books.

Katz, C. (2017). Revisiting minor theory. *Environment and Planning D: Society and Space*, 35(4): 596–599.

Kelley, R. D. G. (1997). *Yo'Mama's Disfunktional!: Fighting the Culture Wars in Urban America*. Boston: Beacon Press.

Kelley, R. D. G. (1998). Check the technique: Black urban culture and the predicament of social science. In N. B. Dirk (Ed.), *In Near Ruins: Cultural Theory at the End of the Century*. 39–66. Minneapolis, MN: University of Minnesota Press.

Kelley, R. D. G. (2002). *Freedom Dreams*. Boston: Beacon Press.

King, J. E. (2017). 2015 AERA Presidential Address Morally Engaged Research/ers Dismantling Epistemological Nihilation in the Age of Impunity. *Educational Researcher*, 46(5): 211–222.

Krueger, P. (2011). Activating memories of endurance: Participatory action research of school safety practices and institutional codes of research ethics. *International Review of Qualitative Research*, 3(4): 411–432.

Kushner, R. (2019, 17 April). Is prison necessary? Ruth Wilson Gilmore might change your mind. *The New York Times*. https://www.nytimes.com/2019/04/17/magazine/prison-abolition-ruth-wilson-gilmore.html

Kumanyika, C. (2020, 20 June). Ruth Wilson Gilmore makes the case for abolition. *Intercepted*. https://theintercept.com/2020/06/10/ruth-wilson-gilmore-makes-the-case-for-abolition/

Lorde, A. (2003). The master's tools will never dismantle the master's house. In R. Lewis & S. Mills (Eds.), *Feminist Postcolonial Theory: A Reader*. 25–27. New York: Routledge.

Loyd, J. M. (2012). Borders, Prisons, and Abolitionist Visions. In J. M. Loyd, M. Michelson, & A. Burridge (Eds.), *Beyond Walls and Cages: Prisons, Borders and Global Crisis*. 1–15. Athens, GA: University of Georgia Press.

Loyd, J. M., & Bonds, A. (2018). Where do Black lives matter? Race, stigma, and place in Milwaukee, Wisconsin. *The Sociological Review*, 66(4): 898–918.

Luttrell, W. (2019). Picturing care: An introduction. *Gender and Education* 31(5): 563–575. https://doi.org/10.1080/09540253.2019.1621502

Mann, S., Nolan, J., & Wellman, B. (2003). Sousveillance: Inventing and using wearable computing devices for data collection in surveillance environments. *Surveillance & Society*, 1(3): 331–355.

Manzo, L., & Brightbill, N. (2007). Towards a participatory ethics. In S. Kindon, R. Pain, & M. Kesby (Eds.), *Participatory Action Research Approaches and Methods: Connecting People, Participation and Place*. 33–40. New York: Routledge.

Martin, D. G. (2007). Bureacratizing ethics: Institutional review boards and participatory research. *ACME: An International Journal for Critical Geographies*, 6(3): 319–328.

Martin, D. G., & Inwood, J. (2012). Subjectivity, power, and the IRB. *The Professional Geographer*, 64(1): 7–15.

McKittrick, K. (2011). On plantations, prisons, and a black sense of place. *Social & Cultural Geography*, 12(8): 947–963.

McKittrick, K. (2014). Mathematics black life. *The Black Scholar*, 44(2): 16–28.

McKittrick, K. (2021). *Dear Science and Other Stories*. Durham, NC: Duke University Press.

McKittrick, K., & Woods, C. (Eds.) (2007). *Black Geographies and the Politics of Place*. Boston: South End Press.

Moraga, C., & Anzaldúa, G. (1983). This bridge called my back: Radical writings by women of color. In M. Anderson & P. Hill Collins (Eds.) (1992), *Race, Class, and Gender: An Anthology*. 20–27. Belmont, California: Wadsworth Publishing Company.

Moynihan, D. (1965). *The Negro Family: The Case for National Action*. Washington, DC: Office of Policy Planning and Research, United States Department of Labor.

Nagar, R. (2014). *Muddying the Waters: Co-Authoring Feminisms across Scholarship and Activism*. Chicago: University of Illinois Press.

Nagar, R. (2019). Hungry Translations: The World Through Radical Vulnerability: The 2017 Antipode RGS-IBG Lecture. *Antipode*, 51(1): 3–24.

Pain, R., Askins, K., Banks, S., Cook, T., Crawford, G., Crookes, L., Derby, S., Heslop, J., Robinson, Y., & Vanderhoven, D. (2015). *Mapping Alternative Impact: Alternative Approaches to Impact from Co-Produced Research*. Durham, UK: Centre for Social Justice and Community Action, Durham University. https://eprints.gla.ac.uk/

Pain, R., & Staeheli, L. (2014). Introduction: Intimacy-geopolitics and violence. *Area*, 46(4): 344–347.

Patel, L. (2016). *Decolonizing Educational Research: From Ownership to Answerability*. London: Routledge.

Pezzullo, P. C. (2020). Resisting carelessness: The care manifesto: The politics of interdependence, by The Care Collective. *Cultural Studies* 36(3): 507–509.

Pulido, L. (2002). Reflections on a white discipline. *The Professional Geographer*, 54: 42–49.

Pulido, L. (2016). Flint, environmental racism, and racial capitalism. *Capitalism Nature Socialism*, 27(3): 1–16. https://doi.org/10.1080/10455752.2016.1213013

Pulido, L. (2017). Geographies of race and ethnicity II: Environmental racism, racial capitalism and state-sanctioned violence. *Progress in Human Geography*, 41(4): 524–533.

Purnell, D. (2021). *Becoming Abolitionists: Police, Protest, and the Pursuit of Freedom*. London: Verso Books.

Reyna Rivarola, A. R., & López, G. R. (2021). Moscas, metiches, and methodologies: Exploring power, subjectivity, and voice when researching the undocumented. *International Journal of Qualitative Studies in Education*, 34(8): 733–745. https://doi.org/10.1080/09518398.2021.1930262.

Ritchie, A. J. (2017). *Invisible No More: Police Violence Against Black Women and Women of Color*. Boston: Beacon Press.

Ritterbusch, A. (2012). Bridging guidelines and practice: Toward a grounded care ethics in youth participatory action research. *The Professional Geographer*, 64(1): 16–24.

Ritterbusch, A. E. (2019). Empathy at knifepoint: The dangers of research and lite pedagogies for social justice movements. *Antipode*, 51(4): 1296–1317. https://doi.org/10.1111/anti.12530

Rolón-Dow, R. (2005). Critical care: A color (full) analysis of care narratives in the schooling experiences of Puerto Rican girls. *American Educational Research Journal*, 42(1): 77–111.

Rothstein, R. (2017). *The Color of Law: A Forgotten History of How Our Government Segregated America*. New York: Liveright Publishing.

Roy, A. (2017). Dis/possessive collectivism: Property and personhood at city's end. *Geoforum*, 80: A1–A11. https://doi.org/10.1016/j.geoforum.2016.12.012

Sabati, S. (2019). Upholding "colonial unknowing" through the IRB: Reframing institutional research ethics. *Qualitative Inquiry*, 25(9–10): 1056–1064.

Sampson, R. J., & Raudenbush, S. W. (2004). Seeing disorder: Neighborhood stigma and the social construction of "broken windows." *Social Psychology Quarterly*, 67(4): 319–342.

Sandwick, T., Fine, M., Greene, A. C., Stoudt, B. G., Torre, M. E., & Patel, L. (2018). Promise and provocation: Humble reflections on critical participatory action research for social policy. *Urban Education*, 53(4): 473–502.

Scannell, R. (2018). *Electric light: Automating the Carceral State during the Quantification of Everything*. PhD thesis. New York: City University of New York. https://academicworks.cuny.edu/gc_etds/2571.

Shabazz, R. (2015). *Spatializing Blackness: Architectures of Confinement and Black Masculinity in Chicago*. Chicago: University of Illinois Press.

Sharpe, C. (2014). Black life, annotated. *The New Inquiry*, 8. https://thenewinquiry.com/black-life-annotated/

Sharpe, C. (2016). *In the Wake: On Blackness and Being*. Durham, NC: Duke University Press.

Sharpe, C. (2021). Black gathering: An assembly in three parts. In S. Anderson & M. O. Wilson (Eds.), *Reconstructions: Architecture and Blackness in America*. 24–29. New York: Museum of Modern Art.

Smith, L. T. (2013). *Decolonizing Methodologies: Research and Indigenous Peoples*. London: Zed Books.

Smith, N. (2001). Global social cleansing: Postliberal revanchism and the export of zero tolerance. *Social Justice*, 28(3), 68–74. https://www.jstor.org/stable/29768095

Spade, D. (2020). *Mutual Aid: Building Solidarity during this Crisis (and the Next)*. London: Verso Books.

Spivak, G. C. (2007). *Can the Subaltern Speak?* Vienna: Verlag Turia & Kant.

Stein, S. (2019). *Capital City: Gentrification and the Real Estate State*. London: Verso Books.

Stoudt, B., Fine, M., & Fox, M. (2011). Growing up policed. *New York Law School Law Review*, 56(4). 1331–1372. https://www.nylslawreview.com/

Stoudt, B. G., Cahill, C., Belmonte, K., Djokovic, S., Lopez, J., Matles, A., Torre, M. E., & X, D. (2016). Participatory action research as youth activism. In J. Conner & S. Rosen (Eds.), *Contemporary Youth Activism: Advancing Social Justice in the U.S.* 327–346. Santa Barbara, CA: Praeger.

Stoudt, B. G., Torre, M. E., Bartley, P., Bissell, E., Bracy, F., Caldwell, H., Dewey, L., Downs, A., Greene, C., & Haldipur, J. (2019). Researching at the community-university borderlands: Using public science to study policing in the South Bronx. *Education Policy Analysis Archives*, 27. https://epaa.asu.edu/index.php/epaa/article/view/2623

Sultana, F. (2007). Reflexivity, positionality and participatory ethics: Negotiating fieldwork dilemmas in international research. *ACME: An International Journal for Critical Geographies*, 6(3): 374–385.

Torre, M. E. (2006). *Beyond the Flat: Intergroup Contact, Intercultural Education and the Potential of Contact Zones for Research, Growth and Development*. Unpublished manuscript.

Torre, M. E., Cahill, C., & Fox, M. (2015). Participatory Action Research in Social Research. In J. D. Wright (Ed.), *International Encyclopedia of the Social & Behavioral Sciences*. 540–544. Amsterdam: Elsevier. https://doi.org/10.1016/B978-0-08-097086-8.10554-9.

Torre, M. E., Fine, M., Alexander, N., Billups, A. B., Blanding, Y., Genao, E., Marboe, E., Salah, T., & Urdang, K. (2008). Participatory action research in the contact zone. In J. Cammarota & M. Fine (Eds.), *Revolutionizing Education: Youth Participatory Action Research in Motion*. 23–44. New York: Routledge.

Torre, M. E., Fine, M., Stoudt, B. G., & Fox, M. (2012). Critical participatory action research as public science. In H. Cooper, P. M. Camic, D. L. Long, A. T. Panter, D. Rindskopf, & K. J. Sher (Eds.), *APA Handbook of Research Methods in Psychology, vol 2: Research Designs: Quantitative, Qualitative, Neuropsychological, and Biological*. 171–184. Washington, DC: American Psychological Association.

Torre, M. E., Stoudt, B., Manoff, E., & Fine, M. (2017). Critical participatory action research on state violence: Bearing wit(h)ness across fault lines of power, privilege, and dispossession. In N. K. Denzin & Y. S. Wadsworth (Eds.), *The SAGE Handbook of Qualitative Research*. 492–515. Thousand Oaks, CA: Sage Publications.

Tuck, E. (2009). Suspending damage: A letter to communities. *Harvard Educational Review*, 79(3): 409–428.

Tuck, E., & Gorlewski, J. (2016). Racist ordering, settler colonialism, and edTPA: A participatory policy analysis. *Educational Policy*, 30(1): 197–217.

Tuck, E., & Guishard, M. (2013). Scientifically based research and settler coloniality: An ethical framework of decolonial participatory action research. In T. Kress, C. Malott, & B. Porfilio (Eds.), *Challenging Status Quo Retrenchment: New Directions in Critical Qualitative Research*. Charlotte, NC: Information Age Publishing.

Tuck, E., & Yang, K. W. (2014). Unbecoming claims: Pedagogies of refusal in qualitative research. *Qualitative Inquiry*, 20(6): 811–818.

Valenzuela, A. (1999). *Subtractive Schooling: U.S.-Mexican Youth and the Politics of Caring*. Albany, NY: SUNY Press.

Vitale, A. S. (2017). *The End of Policing*. London: Verso Books.

Wallace, D., & Wallace, R. (1998). *A Plague on Your Houses: How New York Was Burned Down and National Public Health Crumbled*. London: Verso.

Welfare Warriors Research Collaborative (2010). *A Fabulous Attitude: Low Income LGBTGNC People Surviving and Thriving with Love, Shelter, and Knowledge*. New York: Queers for Economic Justice, Graduate Center of the City University of New York. https://nyf.issuelab.org/

Wilder, C. S. (2000). *A Covenant with Color: Race and Social Power in Brooklyn 1636–1990*. New York: Columbia University Press.

Wilder, C. S. (2014). *Ebony and Ivy: Race, Slavery, and the Troubled History of America's Universities*. New York: Bloomsbury Publishing.

Wilson, J. Q., & Kelling, G. L. (1982). Broken windows. *Atlantic Monthly*, 249(3): 29–38.

Woods, C. A., & Gilmore, R. W. (2017). *Development Arrested: The Blues and Plantation Power in the Mississippi Delta*. London: Verso Books.

3 We Sing the Land

Researching for, with and as Country in North East Arnhem Land, Australia including

Bawaka Country, including L. Burarrwanga, R. Ganambarr, M. Ganambarr-Stubbs, B. Ganambarr, D. Maymuru, S. Wright, K. Lloyd, and S. Suchet-Pearson

Our Songspirals Are Our Land Rights

Yolŋu people in North East Arnhem Land, northern Australia, have been at the forefront of the struggle for recognition of Indigenous sovereignty in Australia for well over half a century. In the 1970s, the first legal case for land rights in Australia was heard when Yolŋu leaders led a legal action against the mining company, Nabalco, who wanted to develop a bauxite mine on their Country (Dunlop, 1995; Morphy, 1983; Morphy & Morphy, 2006; Watson-Verran, Chambers, & the Yolŋu community at Yirrkala, 1989). As we write in our book *Songspirals*, this was a challenge to Western assumptions of superiority and development inherent in the legal fiction of *terra nullius:*

> When the court failed to recognise Yolŋu land rights, it was because they said we didn't have their kind of agriculture or fences or anything that they could see or recognise as using our land. We don't use the land, not in that way; we sing the land so that new trees grow, new plants come, animals flourish. We build our fish traps, we hunt and dig, we nourish our ganguri, our yams, and they nourish us; fire shapes Country, we light the land to renew it. It is sustainable for Yolŋu and everything to live there together. With our songspirals the land renews itself, our songspirals are our land rights.
>
> (Burarrwanga et al., 2019, pp. 258–259)

This chapter discusses our work as an Indigenous-non-Indigenous, more-than-human research collective, focusing on our work sharing songspirals. For Yolŋu people, the beauty and purpose of songs and songspirals are always multilayered. Songspirals are sung by Yolŋu to awaken Country, to make and remake the life-giving connections between people and place. The depth of the spirals is in their meanings, in the patterns, relationships and connections they create again and again. It is infinity, co-becoming with Country, with land, sea, and sky, with all the beings, all the processes, all that is tangible and intangible, that emerge together there. This means that while our work together can be framed as PAR, with a focus on art-based practice including intergenerational and intercultural sharing through song, painting, weaving, and writing together, it is always more. It is more beautiful, more layered, more-than-human, more political, and always more challenging as processes of injustice and recolonisation continue.

We write as the Bawaka Collective led by Bawaka Country. As a knowledgeable, sentient, and caring participant in our work together, Bawaka Country shapes and

DOI: 10.4324/9780429400346-3

guides what we do, the songs we share, and our multilayered more-than-human communications. As it shapes and moulds, so it co-becomes with/as us. We also write as four Yolŋu elders and sisters: Laklak, Ritjilili, Merrkiyawuy, and Banbapuy, and our daughter Djawundil, who live in Yirrkala, North East Arnhem Land. Yirrkala was the first community to talk about treaty and land rights, an icon when talking about land and its power. We are leaders in our community, teachers, and researchers. Banbapuy has taken the lead in this chapter, as the main spokesperson for our family and as an author, artist, weaver, and dedicated teacher. We also write with and as Sarah, Sandie, and Kate, academic human geographers from down south.

Our collaborative work on songspirals has different layers. By focusing on sharing songspirals between generations and cultures, including through workshops and the co-authoring of our book *Songspirals*, we enact participatory action research through our always-in-the-making process of co-becoming that drives our work together. Our coming together over 17 years has been intense, rewarding, and challenging, and has developed through careful thought and much discussion over issues of reciprocity and mutual benefits. Sarah, Kate, and Sandie are sensitive to the challenges, although they do not always meet them. They try to remain aware of, but not paralysed by, the injustices, complexities, grief, and potential for recolonisation that working together entails, as indeed any of their work on the stolen lands where they live does. Together in this chapter, we share some of our learnings; some as a collective and some of it directly from Laklak, Ritjilili, Merrkiyawuy, Banbapuy, and Djawundil. As the Yolŋu collective members have said:

> . . . we are also sharing and learning among ourselves. We cannot let this knowledge fade away. It has been here for so long and it is still here. That is why this book is so important, to pass knowledge down between the generations and to continue the spirals. We are all learning. This book is a journey for us too, a journey of intergenerational learning. We have learnt from our sisters and our mothers. Our mothers have given this to us. Now we need to bring it out from ourselves. And from ourselves to you. It needs to happen now and we want you to walk with us on this journey.
>
> (Burarrwanga et al., 2019, pp. xxvi–xxvii)

In this chapter, we focus on our understandings of research as co-becoming, and discuss some of the pitfalls, silences, and dangers in sharing Yolŋu knowledges in the context of colonial invasion. This includes the importance of understanding art and song in terms of Yolŋu sovereignty and land rights, as well as the central role of place, what we call Country, in PAR through its agency and communication. Yolŋu art, including weaving, painting, and songspirals, should not be taken out of context. These are not things to be admired and collected – they are land rights, they are politics, they are processes and enactments, they are Law and ceremony and sovereignty. Through nurturing and sharing songspirals, Yolŋu nourish Country and it nourishes them; it guides and teaches, it leads our work together. PAR cannot take place with a backdrop of Country and place. Country is active, all beings – human, animal, plant, process, thing, or affect – are constituted through relationships that are constantly re-generated. We then position our research not as participation, as if we could participate in something that sits completely outside us, but as co-becoming. By understanding our process of research and sharing as co-becoming, we aim to centre

Yolŋu cosmologies, in ways that centre co-emergence with and through difference. Through co-becoming, we act, think, and share *as* Country through ongoing, emergent, and relational processes (Bawaka Country et al., 2016).

"This Is My Story and This Is My Painting, Take It with You," He Said

We dedicated our book *Songspirals* to Laklak, Ritjilili, Merrkiyawuy, and Banbapuy's grandfather: their mother's father, their ŋathi, Djulwanbirr Mungurrawuy Yunupiŋu. Mungurrawuy was a spokesperson for Yolŋu people and a Gumatj leader. He shared his artwork with ŋäpaki (non-Indigenous people) who came to Yirrkala to develop the Nabalco bauxite mine but who could not understand. As we say in *Songspirals* (Burarrwanga et al., 2019):

> This book is dedicated to our grandfather, who was such a strong man and who worked so hard to communicate to ŋäpaki what we are explaining with this book, even though English was his fourth or fifth language. Our grandfather gave his paintings to ŋäpaki visitors but got nothing in return. "This is my story and this is my painting, take it with you," he said, sharing the depth of our connection with them. He shared his story, expecting something for it in return, some respect, some recognition of our rights, but he got nothing. Many ŋäpaki couldn't or wouldn't understand. Now maybe they will.
>
> (pp. 259–260)

Well over 50 years of damage and destruction are evidence of the lack of ability of many ŋäpaki to recognise and respect what our grandfather so generously shared with them, the depth of Yolŋu connection with Country. Many ŋäpaki, including politicians, bureaucrats, industry executives, and employees, wouldn't or couldn't understand. In sharing his bark painting he was sharing something much more profound than what is captured by the English term *art* or *artwork*, even though many would have chosen to accept his gift as a mere novelty or an interesting artefact. Through his art, Mungurrawuy was sharing his story, his painting, his song, his Law and sovereignty, his ownership, and his more-than-human connectivities and responsibilities with and as Country.

It is not possible to selectively pick and choose aspects of Yolŋu cosmologies, relationships, and communications, and/or to try and fit them into a Western lens. That is, to take the comforting without the confronting, the cultural without the political, the story without the Law, the art without the ownership. All these dichotomies are meaningless for Yolŋu, and to impose them, even in unwitting or unthinking ways, is to continually recolonise. This is something emphasised by many Indigenous scholars who point out that Indigenous knowledges, whether relating to art or story or environmental practice, never stand separately from assertions of sovereignty, issues of land, or broader relationships of more-than-human responsibility (Langton, 1998; Moreton-Robinson, 2015; Smith, Smith, Wright, Hodge, & Daley, 2020; Todd, 2016; Tuck & Yang, 2012; Watts, 2013).

This is true, even if a researcher's intention is to "speak back" to Anglo-European theorising or engage with different ways of knowing and being. Indigenous Law is never a "trinket," a "sideshow," or a theoretical novelty (Langton, 1998; Moreton-Robinson, 2015; Smith et al., 2020; Todd, 2016; Tuck & Yang, 2012; Watts, 2013). As Vanessa Watts (2013) powerfully describes:

Figure 3.1 Mungurraway with artwork

> These types of historical Indigenous events (i.e. Sky Woman, the Three Sisters) are increasingly becoming not only accepted by Western frameworks of understanding, but sought after in terms of non-oppressive and provocative or interesting interfaces of accessing the real. This traces Indigenous peoples not only as epistemologically distinct but also as a gateway for non-Indigenous thinkers to re-imagine their world. In this, our stories are often distilled to simply that – words, principles, morals to imagine the world and imagine ourselves in the world. In reading stories this way, non-Indigenous peoples also keep control over what agency is and how it is dispersed in the hands of humans.
>
> (p. 26)

Art and song are intimately related to knowledge because Yolŋu language is still sung and spoken and danced today. Yolŋu people still live a Yolŋu way of life, and this means learning about ceremonies, learning to hunt and collect bush food, all the things that were done for thousands and thousands of years and still happen to this day. Land rights is a new story for Yolŋu because, as Banbapuy says:

> We know we belong to the land and the land belongs to us. But we also know we have been primitive people or natives in the eyes of the Australian Constitution and many of its people. Therefore, we need to have the land rights put in place so we are recognised as people, as a nation, and then perhaps a treaty will come.

So, participation or action or research cannot happen separately from land, from Country, and from Yolŋu relationships to people, nonhumans and place. Yolŋu stories and song and art, practices, and relationships, do not describe a separate reality, a separate world, a discrete environment, but bring it into being. Yolŋu songspirals sing the world; Yolŋu stories enact it. Art is Yolŋu Law and relationships and protocols, Yolŋu land and sea rights and sovereignty. To ignore the realities of Yolŋu relationships with land, sea, air, and animals, to ignore the complex co-becomings of Country, to think and act like anything is just theory, is to contribute to ongoing colonising processes. Those processes have a long history and a profound ongoing influence in and as the present. As we describe in our book:

> When the British came, they didn't see, or they ignored or refused to see, the songspirals, the Law, the culture that is here. And they claimed the land. They had only been here for the shortest time and they claimed it. But the land was already claimed. We have boundaries, clan boundaries, we have Law, culture and language. We know which clan belongs to which land.
>
> (Burarrwanga et al., 2019, p. x)

Colonisation is not homogenous nor situated in the distant past, but is an ongoing structure maintained through ongoing practices, knowledges, and legal/economic/political frameworks (Kehaulani Kauanui, 2016; Moreton-Robinson, 2015). This is also true of Indigenous practices and diverse survivances. These too have histories within the present and are multiple and agential. When Banbapuy teaches Garma (two-ways maths) in the classroom or as a form of professional development, she always starts from the history, so people understand where it is coming from, why we are doing that. Banbapuy explains that:

> when we teach the Rom, Yolŋu Law, for example the cleansing ceremony, we teach that it has been done this way for thousands of years. So, when I teach a dance, I have to explain where it is coming from and why we are doing it. We are not doing it for no reason. With feeling and pride, we do the movements. We also talk about mathematics, we start with talking about a point. In Western mathematics, ŋäpaki talk about Plato and how he understood the world, the particular way he talks about a point. In the Yolŋu world, we talk about culture; we talk about the Two Sisters, how when they penetrated the land and out flowed the water, that was the point. Everything begins with the point.

This is not to be toyed with. The stakes are high. It is life and death, for Yolŋu and all beings that co-become together (Larsen & Johnson, 2017). And this is why it is so important to understand the place of Country, the ways that arts-based practice and participation are always more than art. They are our co-becoming. And, in the communications of our collective, we try to make sure we tell the fuller story:

> After that land case a lot of our old people were so saddened by what the court said – that this was not their land – that it killed them. For us, for the new generations, we feel it is time that the stories we are telling in this book must be told. So ŋäpaki can start to understand why we don't want our land to be devastated by destructive mining, digging, farming, putting poison on the land.
>
> (Burarrwanga et al., 2019, p. 259)

Researching as Country: Co-Becoming Not Participating

Co-becoming is our framing of PAR. Rather than any individual agency or group of beings participating in a separate project or process, we see ourselves as always in emergence in entangled, more-than-human, power-laden ways. Here, we discuss the importance of co-becoming as Country in our research before focussing on some of the ways we co-become through our collective work, and how these processes can nurture healing and challenge colonising processes.

Attending to Country

Country is an Australian Aboriginal English term used to refer to specific places or territories that are Aboriginal peoples' homelands (Bawaka Country et al., 2016; Kwaymullina, 2008; Milroy & Milroy, 2008; Rose, 2004; Rose & Australian Heritage Commission, 1996). Country is not a static background to culture, or a stage upon which a range of actors come together to participate in something. Country, and the songspirals that awaken and enliven it, is an intimate part of what it means to know, live, and co-become (Bawaka Country et al., 2016). Indeed, Country anchors Yolŋu people in infinite cycles of kinship, sharing, and responsibility (Bawaka Country et al., 2019; Johnson & Larsen, 2013; Suchet-Pearson, Wright, Lloyd, & Burarrwanga on behalf of Bawaka Country, 2013). As described above, when this is not recognised, respected, or valued, it is devastating, indeed life threatening.

Our more-than-human collective research is encompassed by this understanding of Country. Bawaka Country encompasses all the nonhuman and human beings which emerge together, as kin, to make up Country. For the Bawaka Collective, this includes the eight authors named on this chapter. Kate, Sarah, and Sandie are brought into Bawaka Country through their work on and as Country, and by being placed into Yolŋu kinship systems and attendant responsibilities and obligations (Bawaka Country et al., 2019).

In recognition of the nonhuman centred way in which knowledge co-becomes in our work, we author this chapter and other academic articles as "Bawaka Country including L. Burarrwanga, R. Ganambarr, M. Ganambarr-Stubbs, B. Ganambarr, D. Maymuru, S. Wright, K. Lloyd, S. Suchet-Pearson". This is an effort to recognise the active agency of Country, the way more-than-human agencies shape our collective work. It also recognises that Country is not an undifferentiated mass, but in the case of our academic articles, includes the agencies of specific human beings. Here, Bawaka Country is not a participant in a research agenda, rather Bawaka Country directs our collective work because in life it is the starting point for many of us. As a more-than-human us, we collectively yet contextually shape, enable, and guide our work together.

Attending to Songspirals

To co-become with Country requires an ongoing commitment to paying attention to multiple active agencies and the messages that they send out, a recognition that communication can be nonverbal, and attendant responses are constantly reflected upon. For our recent book, five particular songspirals guide the text:

> Wuymirri, the Whale; Wukun, the Gathering of the Clouds; Guwak, the Messenger Bird; Wititj, the Settling of the Serpent; and Goŋ-gurtha, the Keeper of the Fire. Each of these songspirals reaches out and connects clans and peoples and Country across North East Arnhem Land, across northern Australia, across Australia, to the centre and down south, and north into the currents of the Arafura Sea and to the islands of Melanesia, Polynesia and beyond. They connect us down through the soil and up into Sky Country, into space.
> (Burarrwanga et al., 2019, p. xxiii)

These songspirals presented themselves for the book. They emerged over time, in different ways and different contexts. For example, Wukun was always going to be in the book as a special songspiral that Laklak, Ritjilili, Merrkiywuy, and Banbapuy's Mum shared with Banbapuy when she was ready to pass away in hospital. Wititj emerged as a songspiral in the book in honour of Laklak, whose totem is Wititj, who shares some of her names with the rainbow serpent and because of the snake who was watching when Laklak visited her mother and sister before she was born. Indeed, nothing is separate from place, an agential and sentient place, which directly shapes what can be known and what cannot. Country shows others who we are, what we value, and where we come from. Embracing this agency in PAR is critical for revealing, challenging, refusing, and reshaping colonising processes, which all too often denigrate, devalue, or ignore critical connectivities and responsibilities:

> . . . songspirals link us through time and place, and bring those times and places together, the land and the people. The spiral is infinite, it is how kinship spirals through the generations, from before and into the future. It is all connected.
> (Burarrwanga et al., 2019, p. 261)

Working Together, Learning Together, Meeting

Our work together is always situated within a political imperative to reshape power relationships through sharing, nurturing, and healing. In framing our sharing approaches, we draw on the way Yolŋu children learn. This means sharing through showing. Not theories, not just talking, but showing people. This is what we do with ŋäpaki, what we have done with our book *Songspirals:* brought the reader with us, so they co-become through our footsteps.

Through our book, ŋäpaki can see that there are women who are willing to help, who have knowledge, coming from a ŋäpaki university (Sarah, Sandie and Kate) and a Yolŋu university (Laklak, Ritjilili, Merrkiyawuy, Banbapuy and Djawundil) – the knowledges meeting to help people understand. Through our book ŋäpaki can see the way we work together to help ŋäpaki and Yolŋu see things differently. It is a process of meeting – two knowledges meeting together, working together, learning together. The ŋäpaki researchers are learning from the Yolŋu researchers, and the Yolŋu researchers are learning from the ŋäpaki researchers. It is making a connection of understanding. It is like, you can do this by doing this. And it is giving other people an idea of how people can work together.

When we first met in 2006, Laklak, Ritjilili, Merrkiyawuy, Banbapuy, and Djawundil say they grabbed the chance to work with Sarah, Sandie, and Kate because they are ŋäpaki academics who had knowledge that could help them. Sarah, Sandie, and Kate have that language to approach ŋäpaki and the hierarchy, and Laklak, Ritjilili, Merrkiyawuy, Banbapuy, and Djawundil have the knowledge of Yolŋu and their hierarchy. Sarah, Sandie, and Kate know their ŋäpaki culture and their language; they know what to do. Laklak, Ritjilili, Merrkiyawuy, Banbapuy, and Djawundil know Yolŋu language, culture, and what to do; how to approach people. Ŋäpaki came to Australia on the boat by sea. So, there is a knowledge coming in. And there is the knowledge of Yolŋu, from the land. Here they are meeting.

And we must practice what we say. When we talk about songs, we must practice those things. Not just practice those things but do it, show people how it is connected and how they learn from us. We have written a book, but we are also running workshops, bringing Yolŋu women together to keen the songspirals, to share their knowledge and bring the songspirals and Country to life. Knowing what we teach, what we do, what we can deliver, what we understand. As we have said:

> That's the thing with Yolŋu and culture, everything is a whole, everything is one. We do our own djäma (work), for the self, but really we are one big living thing. And that's why everyone goes through that same, sorrow, crying. Together.
>
> (Bawaka Country et al., 2019, p. 3)

Understanding, Limiting, Refusing

In our collective work, the Yolŋu sisters try to share these understandings with ŋäpaki but there are limits to ŋäpaki understanding, as there should be. We must all, after all, speak from our own place, from our relationships. We must be aware of, and accountable for, our different positions within unjust systems (Bawaka Country et al., 2018; Noxolo, Raghuram, & Madge, 2008; Rose, 2004; Smith et al., 2020). Research and participation, art-based methods, and relationships, all take place within knowledge systems "contaminated by colonialism and racism" (Battiste, 2008, p. 503).

Understanding does not come easily. Ŋäpaki admire much of what Banbapuy, Merrkiyawuy, Ritjilili, Laklak, and Djawundil share yet they do not always understand, because the language used is a higher language; it is the sign language, the maths, the ancient language too. All ŋäpaki understand, it often seems, is the culture bit; that Yolŋu have a culture. And with respect they sit and watch but they don't get anything more. Even if ŋäpaki men are taken into men's business they often don't have the understanding; it doesn't necessarily have the deep meaning for them. They do it out of respect but not understanding.

So, there are limits about what can be understood, and important positionalities and histories to attend to. As *we* talk about co-becoming as a collective, this does not mean *we* are all the same. We come into our collective with different histories, different positionalities, different relationships to Yolŋu Law and sovereignty. Laklak, Ritjilili, Merrkiyawuy, Banbapuy, and Djawundil share knowledge as an assertion of sovereignty. In doing so, they are part of a longstanding Yolŋu movement, which has seen diverse responses and refusals to colonialism, including through land rights, bilingual education, and the homelands movement (Dunlop, 1995; Morphy, 1983; Morphy & Morphy, 2006; Watson-Verran et al., 1989). Kate, Sarah, and Sandie are non-Indigenous, situated very differently both within the collective and more broadly within systems of injustice. As such, their sharings, co-becomings, doings, and knowings are always underpinned by a range of tensions and complicities, including the ways Sarah, Sandie, and Kate are privileged and enabled by the ongoing theft of land of the Aboriginal Countries in which they live, thousands of kilometres to the south, on Gumbaynggirr and Dharug lands.

And the *we* in this article is more-than-human too. We are a collective that emerges through more-than-human co-becomings in ways that go beyond human understanding. No human can know all that a wind knows, or know in the same way a whale knows, or have the same knowledges as the sand. And yet, the relationships hold; the knowledge is in the songspirals. The co-becoming of whales and people happen in deep ways, real ways, as they journey together, in the rich and intimate connections felt between whale and hunter, whale and mother. And it is in the co-becoming of the messages sent, the knowledge made at Bawaka when we run across hot sand in the season of Rrarrandharr. These co-becomings are beautiful and real in ways that go beyond what can be understood or felt – or even be real in many Western ways.

So, as we share what we do share, although we do not share everything, there are incommensurabilities that just can't be translated (Hunt, 2014; Rubis & Theriault, 2020). Banbapuy asks herself if ŋäpaki understand as she shares, or whether her sharing of knowledge or her translations are a waste of time. Maybe, she thinks, we researchers need to work together to help ŋäpaki understand. We learn and we unlearn.

Understanding different positionalities, attending to incommensurabilities and limits is fundamental to co-becoming well, but this is not something to necessarily overcome through more and more knowledge. This is not the space to replicate a colonising acquisition. Understandings and relationships come from the foundation of your place and culture, in all its beauty and its violence.

Power dynamics are not displaced by good intentions either, or through good relationships (though these are important), or with more and more knowledge. The ethics of our uneasy spaces of diverse co-becomings, as we try to work in decolonising ways, are far from clear (Barker & Pickerill, 2020). In co-becoming together, our relationships are always in emergence, always connected to different histories and presents, and must always be attended to (Coddington, 2017; Lloyd, Wright, Suchet-Pearson, Burarrwanga, & Bawaka Country, 2012). There is no easy, static, or achievable ethical position but rather an "indefinite uncertainty" (Coddington, 2017 p. 318), that requires "response-ability as a dynamic, contextual and ongoing way of working" (Bawaka Country et al., 2019, p. 12).

If we understand our place, if we come from place, we can have good relationships. If we know our own foundation, our own culture, we know where we stand. As Banbapuy explains, together we share the histories of our nations. We do our collective work because we know our different stories, we share because we are telling and educating people about the two worlds that can be shared and knowledges that can work together and be researched to help our next young generation build bridges for their life, to live on. We share our stories and ideas together as people, mothers, grandmothers. We understand who we are: we have given life the same way, no different, we are the first teachers.

That is why we have a research agreement, one that evolves and charts our history. We update our story each time we revisit it, and we use it for remembering, communicating with each other and with others. Created in July 2008 and updated in 2009, 2010, 2011,

2014, 2017, 2019 and 2023, it outlines our partnership, our history together, and past, current, and future projects. It is used in the Bawaka Cultural Experiences Business Plan submitted to University Ethics Committees and is a document to help us achieve outcomes and processes that are useful for everyone. This document is representative of how our arts-based methods create diverse outputs, including books, academic journal articles, weaving workshops, and educational handbooks, which are powerful mechanisms for sharing Yolŋu knowledge both through intergenerational exchange and to ŋäpaki audiences as well. We also use it as a check point to try and avoid falling into deep colonising traps, including how we communicate to each other and others.

And it is important to understand that the very agenda setting is potentially fraught. We are very wary of "participating" in structures set by others. Often these structures are colonising, and participating in them, even if it is to "speak back," can sometimes mean feeding those structures (Coulthard, 2014; Simpson, 2007; Wright, 2018). This is true even if it is about participating in a form of action research that is set with the best of intentions. In our collective, we try to set our agenda through our processes of more-than-human co-becoming. It is set by Country, by songs, by whales and winds and waters, and our relationships with them. Thinking that academic researchers, whether Indigenous or not, are the only ones who can set agendas, or that only humans can or do, is to follow and reinforce colonising logics (Chiew, 2014; Meissner, 2017). Rather than feed these logics, we try to feed those things that are generative and nurturing for ourselves, for Yolŋu Law, for our songspirals, and we do this through co-becoming Country (Coulthard, 2014; Simpson, 2007). We don't just speak back to the things that oppress, we actively centre Yolŋu Law, Yolŋu co-becoming. As we say in our book, it is a message: "This book is important and powerful because it comes out of Yolŋu minds, Yolŋu hearts, Yolŋu mouths. It is us, speaking for ourselves" (Burarrwanga et al., 2019, p. x).

Conclusion

Issues of power are at the heart of PAR and PAR is situated and enmeshed in relationships of power. The Bawaka Collective has emerged through our more-than-human efforts to nurture connectivities and responsibilities and to shift understandings and relationships through generous processes of sharing. Based on Yolŋu ontological understandings that there is no separate "thing" that individual beings can participate in (equally or otherwise), we frame our work as co-becoming. This is an emergent, contextualised coming together of more-than-human agencies, which is encompassed in the concept and actuality of Country, and which opens itself to attending, respecting and responding to Country's guidance. Our work sharing songspirals, including through this paper, is an exemplar of Country's authorship and authority. It is also a reminder that songs, stories, art, and culture cannot be neatly compartmentalised, romanticised, or tokenised. They are all power-laden and power-inducing, and our work together in this space actively works to speak back to the challenges and injustice of ongoing colonising processes.

This work is always a challenge. Even talking about colonising process risks recentring the very power dynamics we seek to shift. Through our work together – from the inception of ideas and projects, the experiencing, sharing, story-ing, the drawing together of information and sharing through workshops, lectures or co-authoring published pieces – we co-become in complex, differentiated, more-than-human ways. There is no simple template for how this unfolds; it is not a linear process, but is deeply contextual and always in emergence.

We base our work on Yolŋu control, the Yolŋu authority to lead and shape and refuse. Through our more-than-human co-becoming, we centre Yolŋu, Bawaka cosmologies in our work as a form of challenging, undermining, and refusing the disciplines of colonial structures. As we do so, we recognise this is not simple nor always easy. Mostly, it is not even possible in any pure way. For as we co-become together, we do so in and from our differential places, in ways always shot through with colonising logics and power relations. There is not a them nor us, not a humans separate from Country, not Indigenous as separate from non-Indigenous. Yet neither is it an undifferentiated mess or a hippy wholeness; it is a co-becoming with limits and boundaries and tensions and, we hope, a respect for difference as we co-become, not despite these issues, but with them, through them. Sometimes the power to decide and control shifts, sometimes on purpose, with respect and with discussion – and sometimes not. We also recognise that, through our collective co-becoming, it is not necessarily useful to weigh up a balance of power between Yolŋu and ŋäpaki. Our collective *we* is always in emergence, always checking in with its more-than-human self/ves to ensure we are sharing, and not sharing, with respect, generosity, and care as far as we can.

> We share songspirals with you and we ask that you treat them with respect. Respecting the knowledge means not writing about things that you don't understand, not putting things into your own words. The words in this book are our knowledge, our property. You can talk about it, but don't think you can become the authority on it. You can use our words for reflection. You can talk about your own experiences and think about how to take lessons from our book into your life. You need to honour the context of our songspirals, acknowledge the layers of our knowledge. You can talk about the very top layer but you need to be respectful and aware of the limits of what we are sharing and what you in turn can share.
>
> (Burarrwanga et al., 2019, p. xxv–xxvi)

References

Ahenakew, C., De Oliveira Andreotti, V., Cooper, G., & Hireme, H. (2014). Beyond epistemic provincialism: De-provincializing indigenous resistance. *AlterNative*, 10(3): 216–231. https://doi.org/10.1177%2F117718011401000302.

Barker, A. J., & Pickerill, J. (2020). Doings with the land and sea: Decolonising geographies, indigeneity, and enacting place-agency. *Progress in Human Geography*, 44(4): 640–662. https://doi.org/10.1177%2F0309132519839863.

Battiste, M. (2008). Research ethics for protecting indigenous knowledge and heritage: Institutional and researcher responsibilities. In N. K. Denzin, Y. S. Lincoln, & T. L. Smith (Eds.), *Handbook of Critical and Indigenous Methodologies*. 497–510. Thousand Oaks, CA: Sage Publications, Inc.

Bawaka Country including Burarrwanga, L., Ganambarr, R., Ganambarr-Stubbs, M., Ganambarr, B., Maymuru, D., Suchet-Pearson, S., Wright, S., Lloyd, K., Tofa, M., & Daley, L. (2018). *Intercultural Communication Handbook*. http://bawakacollective.com/handbook/.

Bawaka Country, Lloyd, K., Suchet-Pearson, S., Wright, S., Burarrwanga, L., Ganambarr, R., Ganambarr-Stubbs, M., Ganambarr, B., & Maymuru, D. (2016). Morrku Mangawu— knowledge on the land: Mobilising Yolŋu mathematics from Bawaka, North East Arnhem Land, to reveal the situatedness of all knowledges. *Humanities*, 5(3): 61–74. https://doi.org/10.3390/h5030061.

Bawaka Country, Suchet-Pearson, S., Wright, S., Lloyd, K., Tofa, M., Sweeney, J., Burarrwanga, L., Ganambarr, R., Ganambarr-Stubbs, M., Ganambarr, B., & Maymuru, D. (2019). Goŋ

Gurtha: Enacting response-abilities as situated co-becoming. *Environment and Planning D: Society and Space*, 37(4): 682–702. https://doi.org/10.1177/0263775818799749.

Bawaka Country, Wright, S., Suchet-Pearson, S., Lloyd, K., Burarrwanga, L., Ganambarr, R., Ganambarr-Stubbs, M., Ganambarr, B., Maymuru, D., & Sweeney, J. (2016). Co-becoming Bawaka: Towards a relational understanding of place/space. *Progress in Human Geography*, 40 (4): 455–475. https://doi.org/10.1177%2F0309132515589437.

Burarrwanga, L., Ganambarr, R., Ganambarr-Stubbs, M., Ganambarr, B., Maymuru, D., Wright, S., Suchet-Pearson, S., & Lloyd, K. (2019). *Songspirals: Sharing Women's Wisdom of Country through Songlines*. Crows Nest: Allen & Unwin.

Chiew, F. (2014). Posthuman ethics with Cary Wolfe and Karen Barad: Animal compassion as trans-species entanglement. *Theory, Culture & Society*, 31(4): 51–69. https://doi.org/10.1177/0263276413508449.

Coddington, K. (2017). Voice under scrutiny: Feminist methods, anticolonial responses, and new methodological tools. *The Professional Geographer*, 69(2): 314–320. https://doi.org/10.1080/00330124.2016.1208512.

Coulthard, G. S. (2014). *Red Skin, White Masks: Rejecting the Colonial Politics of Recognition*. Minneapolis, MN: University of Minnesota Press.

Dunlop, I. (Director). (1995). *Pain for this Land*. Australia: National Film and Sound Archive of Australia.

Hunt, S. (2014). Ontologies of indigeneity: The politics of embodying a concept. *Cultural Geographies*, 21(1): 27–32. https://doi.org/10.1177%2F1474474013500226.

Johnson, J. T., & Larsen, S. C. (2013). *A Deeper Sense of Place: Stories and Journeys of Collaboration in Indigenous Research*. Corvallis, OR: Oregon State University Press.

Kehaulani Kauanui, J. (2016). A structure, not an event: Settler colonialism and enduring Indigeneity. *Lateral: Journal of the Cultural Studies Association*, 5(1). https://csalateral.org/.

Kwaymullina, A. (2008). Introduction: A land of many countries. In S. Morgan, T. Mia, & B. Kwaymullina (Eds.), *Heartsick for Country: Stories of Love, Spirit and Creation*. 6–21. North Fremantle: Fremantle Press.

Langton, M. (1998). *Burning Questions: Emerging Environmental Issues for Indigenous Peoples in Northern Australia*. Centre for Indigenous Natural and Cultural Resource Management, Northern Territory University.

Larsen, S. C., & Johnson, J. T. (2017). *Being Together in Place: Indigenous Coexistence in a More than Human World*. Minneapolis, MN: University of Minnesota Press.

Lloyd, K., Suchet-Pearson, S., Wright, S., Tofa, M., Rowland, C., Burarrwanga, L., Ganambarr, R., Ganambarr, M., Ganambarr, B., & Maymuru, D. (2015). Transforming tourists and "culturalising commerce": Indigenous tourism at Bawaka in Northern Australia. *International Indigenous Policy Journal*, 6(4): 1–21. https://doi.org/10.18584/iipj.2015.6.4.6.

Lloyd, K., Wright, S. L., Suchet-Pearson, S., Burarrwanga, L., & Bawaka Country (2012). Reframing development through collaboration: Towards a relational ontology of connection in Bawaka, North East Arnhem Land. *Third World Quarterly*, 33, 1075–1094. https://doi.org/10.1080/01436597.2012.681496.

Marika, W., as told to Isaacs, J. (1995). *Wandjuk Marika: Life Story*. St. Lucia: University of Queensland Press.

Meissner, H. (2017). Politics as encounter and response-ability: Learning to converse with enigmatic others. *Revista Estudos Feministas*, 25(2): 935–944. https://doi.org/10.1590/1806-9584.2017.v25n2p935.

Milroy, G. I., & Milroy, J. (2008). Different ways of knowing: Trees are our family too. In B. Kwaymullina, T. Mia, & S. Morgan (Eds.), *Heartsick for Country*. 23–46. Freemantle: Freemantle Press.

Moreton-Robinson, A. (2015). *The White Possessive: Property, Power, and Indigenous Sovereignty*. Minneapolis, MN: University of Minnesota Press.

Morphy, H. (1983). Now you understand: An analysis of the way Yolngu have used sacred knowledge to retain their autonomy. In N. Peterson & M. Langton (Eds.), *Aborigines, Land and Land Rights*. 110–133. Canberra: Australian Institute of Aboriginal Studies.

Morphy, H., & Morphy, F. (2006). Tasting the waters: Discriminating identities in the waters of Blue Mud Bay. *Journal of Material Culture*, 11(1–2): 67–85. https://doi.org/10.1177%2F1359183506063012.

Noxolo, P., Raghuram, P., & Madge, C. (2008). 'Geography is pregnant' and 'geography's milk is flowing': Metaphors for a postcolonial discipline? *Environment and Planning D: Society and Space*, 26(1): 146–168. https://doi.org/10.1068%2Fd81j.

Rose, D. B. (2004). *Reports from a Wild Country: Ethics for Decolonisation*. Kensington: UNSW Press.

Rose, D. B., & Australian Heritage Commission (1996). *Nourishing Terrains: Australian Aboriginal Views of Landscape and Wilderness*. Canberra: Australian Heritage Commission.

Rubis, J. M., & Theriault, N. (2020). Concealing protocols: Conservation, Indigenous survivance, and the dilemmas of visibility. *Social & Cultural Geography*, 21(7): 962–984. https://doi.org/10.1080/14649365.2019.1574882.

Sidaway, J. D., Woon, C. Y., & Jacobs, J. M. (2014). Planetary postcolonialism. *Singapore Journal of Tropical Geography*, 35(1): 4–21. https://doi.org/10.1111/sjtg.12049.

Simpson, A. (2007). On ethnographic refusal: Indigeneity, 'voice' and colonial citizenship. *Junctures: The Journal for Thematic Dialogue*, 9: 67–80. https://junctures.org/.

Smith, A. S., Smith, N., Wright, S., Hodge, P., & Daley, L. (2020). Yandaarra is living protocol. *Social & Cultural Geography*, 21(7): 940–961. https://doi.org/10.1080/14649365.2018.1508740.

Snelgrove, C., Dhamoon, R., & Corntassel, J. (2014). Unsettling settler colonialism: The discourse and politics of settlers, and solidarity with Indigenous nations. *Decolonization: Indigeneity, Education & Society*, 3(2): 1–32. https://jps.library.utoronto.ca/index.php/des.

Strakosch, E. & Macoun, A. (2012). The vanishing endpoint of settler colonialism. *Arena Journal, 37/38*: 40–62. https://arena.org.au/category/arena-journal/.

Suchet-Pearson, S., Wright, S., Lloyd, K., Burarrwanga, L., on behalf of Bawaka Country (2013). Caring as country: Towards an ontology of co-becoming in natural resource management. *Asia Pacific Viewpoint*, 54(2): 185–197. https://doi.org/10.1111/apv.12018.

Todd, Z. (2016). An Indigenous feminist's take on the ontological turn: 'Ontology' is just another word for colonialism. *Journal of Historical Sociology*, 29(1): 4–22. https://doi.org/10.1111/johs.12124.

Tuck, E., & Yang, K. W. (2012). Decolonization is not a metaphor. *Decolonization: Indigeneity, Education & Society*, 1(1): 1–40. https://jps.library.utoronto.ca/index.php/des.

Watson-Verran, H., Chambers, D., & the Yolŋu community at Yirrkala (1989). *Singing the Land, Signing the Land: A Portfolio of Exhibits*. Geelong, Australia: Deakin University Press.

Watts V. (2013). Indigenous place-thought and agency amongst humans and non-humans (First Woman and Sky Woman go on a European world tour!). *Decolonization: Indigeneity, Education & Society* 2(1): 20–34. https://jps.library.utoronto.ca/index.php/des.

Wright, S. (2018). When dialogue means refusal. *Dialogues in Human Geography*, 8(2): 128–132. https://doi.org/10.1177%2F2043820618780570.

4 Radical Imaginings

Queering the Politics and Praxis of Participatory Arts-Based Research

John Marnell

> The function of art is to do more than tell it like it is—it is to imagine what is possible.
> – bell hooks

> Queer futures are being shaped every time we create imaginaries of living otherwise.
> – Miguel Á. López

The epigraphs above – the first, a call for social and political transformation through creative border crossings, and the second, a tribute to Giuseppe Campuzano, a *trasvesti* artist, performer and researcher – go to the heart of my politics and praxis.[1] As a queer scholar engaged in participatory arts-based research (ABR), I recognise the potential of radical imaginings to challenge dominant ontological and epistemological paradigms, to disrupt normative social orderings, and to generate alternative futures. In my work with LGBTQ+ migrants, refugees, and asylum seekers,[2] I use various creative research methods to investigate complex lived realities and to support community-driven responses to inequality. My research is deliberately and unapologetically queer, both in subject matter (the intersection of sexuality, gender, and mobility) and orientation (how I employ theory and method). Put another way, I seek to *queer* how I do research as much as I wish to study topics that might be considered queer.

While there is a growing body of literature charting the shared genealogy of feminist and participatory methodologies (e.g., Singh, Richmond, & Burnes, 2013; Frisby, Maguire, & Reid, 2009), far less attention has been afforded to overlaps between queer theory and participatory action research (PAR). This strikes me as peculiar given their many commonalities: both reject positivism and claims to universal truth, both embrace fluid and unstable subject positions, and both value the affective and corporal dimensions of human experience. Perhaps the most striking overlap is their preoccupation with knowledge politics, with both questioning the limits and ramifications of "proper" research.

This chapter explores some of these convergences by reflecting on a multimodal ABR intervention with LGBTQ+ migrants in Johannesburg, South Africa. My initial objective was to work with participants to analyse their relationships to and movements through urban space. However, as the workshop progressed, and as participants became more comfortable expressing themselves creatively, the project changed direction. Participants began to share a wider range of experiences and to experiment with different representational forms. Rather than offer sanitised accounts of their lives, participants chose to honour the fluid, messy, and at times contradictory aspects of their social worlds. I argue that the richness of these creative outputs, coupled with participants' favourable response to the workshop process, testifies to the benefits of a queer-participatory approach.

DOI: 10.4324/9780429400346-4

In reflecting on this project, I hope to elucidate how queer theory inflects all aspects of my ABR/PAR praxis. Researchers interested in creative and/or participatory methods can draw valuable lessons from queer theory yet are often put off by its reputation for abstruseness. In reality, a queer perspective can reveal exciting points of departure for retheorising normative assumptions within all modes of research, including those geared towards social justice. Scholars of all disciplines would benefit from a more nuanced understanding of how sexual and gender norms permeate every human interaction – including research encounters – and how these norms shape the production, interpretation, and circulation of meanings.

Thinking Queerly: Charting a Slippery Lineage

Queer theory resists any one definition. It is best understood as a diverse and unruly body of scholarship that "challenges the taken-for-granted assumptions that serve as foundations to human interactions" (Acosta, 2018, p. 407). Drawing heavily on post-structuralism, queer theory disrupts sexual and gender binaries by tracking their formation in different spatial and temporal locations. In a queer reading, subjectivity is a constellation of multiple unstable positions whose meanings are produced, maintained, and often obscured from criticism by various discourses, structures, and institutions.

By placing social categories within their historical contexts, queer theorists seek to rupture heterosexuality's status as the only normal expression of human desire. In *Gender Trouble*, Judith Butler (1990) exposes how the "heterosexual matrix" upholds the fallacy of biological determinism by rewarding certain behaviours and relationships. For Butler, gender and sexuality are made legible through the compulsory enactment of socially approved scripts. The repeated, stylised performance of these scripts produces the illusion of coherency and naturalness. Incessant policing of gender and sexuality transforms these supposedly immutable categories into "instruments of regulatory regimes" (Butler, 1993, p. 308).

Queer theory provides a framework for challenging heteropatriarchal supremacy by uncovering how social meanings are coded and replicated. In exposing cleavages in the internal workings of identity categories, and in celebrating expressions of gender and sexuality that disrupt normative expectations, queer theory foregrounds modes of resistance that are overlooked in reformist political agendas. But to assume that queer theory is relevant only to gender and sexuality is to miss the point: by exposing the instabilities of all subject positions, queer theory calls into question the "natural" principles according to which society is organised.

Although increasingly recognised as a standalone discipline, queer studies continues to be viewed with suspicion by many academics. Sceptics point to its literary and linguistic roots as its primary weakness, arguing that this signifies detachment from material realities. Others suggest that queer theory's rejection of stable categories closes it off to any meaningful examination of subjectivity (Gamson, 2000). More recently, scholars in the Global South have questioned queer theory's ability to transcend the Euro-American context from which it emerged (Tellis & Bala, 2015). The neocolonial legacies embedded within global queer culture have been strongly critiqued by African scholars, who argue that Western researchers often misinterpret or erase local nuances (e.g., Ekine & Abbas, 2013; Nyeck & Epprecht, 2013). Yet, increasing numbers of African scholars are refashioning queer theory to align with their own needs and interests (e.g., Matebeni, Monro, & Reddy, 2018; Macharia,

2015). Stella Nyanzi (2014) proudly describes her work as queer, using the term to interrogate dissidences within indigenous cultural practices. She rejects the use of Western queer theorists as default anchor points and urges scholars to "reclaim . . . African modes of blending, bending and breaking" (p. 67).

I share these debates not as a distraction from the topic at hand but as a reminder that queer studies is dynamic and amorphous; it remains a site of friction, speculation, and invention. I also choose – as a white Australian who lives and works in South Africa – to foreground debates emanating from the African continent.[3] This means being conscious of tensions circulating around *queer*, both as an identity category and an epistemological framework, and acknowledging that it is a term not fully recognised or embraced. However, these contestations do not diminish the potential of queer theory within Global South contexts, or its possible applications within ABR/PAR. My research praxis has benefited from, and will no doubt continue to benefit from, sustained engagement with queer scholarship. Like Nyanzi (2014), I use the term to invoke complication, disruption, and the many ways in which knowledge production can be troubled. Queer theory's critiques of conventional academic tools are pertinent for those employing an ABR/PAR methodology. As Calogero Giametta (2018) notes, queer theory provides a valuable entry point for interrogating what it means "to be 'participatory' as well as critiquing and transforming rigid research terminology and practices that reinforce the social research canon" (p. 879).

The (Im)possibility of a Queer Method

The inkling that queer theory has something useful to offer research methodologies has been around for a while. Pioneering queer theorists were quick to point out the inadequacies of traditional research paradigms and their complicity in reifying heteronormativity. The need for new modes of knowing is epitomised by Michael Warner's (1993) oft-quoted rallying cry to disrupt academic orthodoxies: "'queer' gets a critical edge by defining itself against the normal rather than the heterosexual, and normal includes business in the academy" (p. xxvi).

Early attempts at doing "business in the academy" differently embraced pastiche, with disparate methodologies blended and repurposed. One of the most famous examples is Jack Halberstam's (1998) "scavenger methodology":

> [It] uses different methods to collect and produce information on subjects who have been deliberately or accidentally excluded from traditional studies of human behaviour. The queer methodology attempts to combine methods that are often cast as being at odds with each other.
>
> (p. 13)

Halberstam complicates the logic of traditional research by advocating for flexible, responsive, and interdisciplinary approaches. This is most evident in his rejection of a singular research perspective – that is, the objective, distant, and reliable observer.

José Esteban Muñoz (1996) raises similar concerns, arguing that traditional forms of research are incapable of capturing the intricacies of queer subjectivities:

> Instead of being clearly available as visible evidence, queerness has instead existed as innuendo, gossip, fleeting moments, and performances that are meant to be

> interacted with by those within its epistemological sphere—while evaporating at the touch of those who would eliminate queer possibility.
>
> (p. 6)

By centring obscure and often furtive scraps of data, Muñoz suggests that queer research is more about perspective and sensibility than subjects or tools.

Despite flagging serious limitations in conventional methodologies, these critiques have little to say on how to actually *do* queer research. Halberstam (1998) and Muñoz (1996) emphasise the unpredictable mesh of possibilities that can emanate from a rupture in disciplinary boundaries, yet they shy away from articulating a specific research framework. Their insistence on repurposing methods from diverse disciplines or reading data "queerly" can seem worryingly abstract for those engaged in empirical research. But this absence of a clearly defined methodology is not a failing of early queer scholarship. Rather, it speaks to a larger tension, specifically the difficulty of marrying queer theory's rejection of unifying concepts with the disciplinary coherency demanded by most research paradigms (Green, 2002).

In the last decade, scholars have begun asking what it means to employ a queer praxis beyond cultural studies. Convinced that queer research is more than a simple application of theory, Kath Browne and Catherine Nash (2016) call for queerness to intersect with "those sets of logical organising principles that link our ontological and epistemological perspectives with the actual methods we use to gather data" (p. 2). They go on to suggest that a queer method is "any form of research positioned within conceptual frameworks that highlight the instability of taken-for-granted meanings and resulting power relations" (p. 4).

Taking their cue from Browne and Nash, scholars in fields as diverse as education (McWilliams, 2016), family sciences (Fish & Russell, 2018), geography (Detamore, 2016), and sociology (Taylor, 2016) show how queer approaches can problematise received analytical categories and destabilise the subject-object dichotomy on which empiricism is predicated. Scholars have also looked to queer theory to reinvigorate established methods, such as oral history (Johnson, 2016) and ethnography (Valocchi, 2005), and to overhaul conventional fieldwork practices (Di Feliciantonio, Gadelha, & DasGupta, 2017).

While distinct in their applications of queer theory, each of these contributions rejects the "individualism, self-congratulatory nature and liberal understandings of both positivist social science and other projects that assume united and coherent subjects and objects" (Dahl, 2016, p. 145). They also share a belief in the radical possibilities of thinking and doing research differently, of reimagining what it means to craft knowledge *with* and *for* those under investigation. This extends to transforming the relational terrain of research engagements (Fields, 2016), embracing the affective and corporal dimensions of knowing (Boyce, Engrebretsen, & Posocco, 2018) and honouring the tensions, fissures, gaps, and inconsistencies inherent in all modes of research (Dadas, 2016). There is also a push to recognise "failure" as an integral component of queer research (Khan & Marnell, 2022; Halberstam, 2011). Scholars like Heather McLean (2018) argue that experimentation, even when it does not lead to "successful" outputs, allows for more creative, cooperative, and chaotic forms of knowledge to emerge.

Cutting across these various interpretations is the notion of a queer orientation. In conceptualising a queer phenomenology, Sara Ahmed (2006) explains how our experience of an object is shaped by our arrival at it – in other words, "how we turn toward that object" (p. 2). Her thinking can be extended to research methodologies, specifically how we might turn toward knowledge production in ways that produce something distinctly queer. Ahmed's work encourages researchers to reconsider how we situate ourselves, with a view of eliciting new modes of seeing, experiencing, doing, and feeling. It is in this act of reorienting – not just in relation to the object of study but also toward the "truth-finding mechanisms" of empiricism (Giametta, 2018, p. 871) – that we can disrupt the normative foundations of research. A queer methodology can thus be understood as one that makes "space for what is" (Crosby et al., 2011, p. 144), that questions subject boundaries and dominant paradigms (McCann, 2016) and that embraces multiplicities, misalignments, and the fertile interstices between theory, method, and lived experience (Brim & Ghaziani, 2016).

In many respects, this conceptualisation of a queer methodology mirrors the preoccupations of ABR/PAR. This is most apparent in their mutual commitment to prioritising identities, voices, experiences, and perspectives that are routinely disregarded. Yet, these similarities should not be mistaken for congruence; it would be incorrect to label all ABR/PAR as queer or to assume that all queer methods value expressive or embodied forms of knowledge. In attempting to queer my research, I draw on ABR's efforts to combine "rigorous qualities of inquiry and artistic means" (Lenette, 2019, p. 32), while simultaneously questioning normative assumptions lurking within its modes and practices. My work is firmly geared towards action, merging queer theory's impetus for epistemological disruption with ABR/PAR's focus on social transformation. Action here is not understood as large-scale shifts in policy, or even as material benefits for individuals, but rather as efforts to generate new ways of thinking, doing, and being. It is in this potential to challenge social, epistemological, and pedagogical conventions that I see my politics and praxis coalescing in productive ways. For me, a queer-participatory approach serves as a reminder "that there are possibilities for things to be otherwise" (Foster, 2016, p. 1).

Creative Resistance: Art as Research, and Research as Art

Inspired by popular education movements, PAR aims to democratise research processes by employing "sequential reflection and action carried out with and by local people rather than on them" (Cornwall & Jewkes, 1995, p. 1667). It is this commitment to inclusion, collaboration, and action that defines PAR, rather than the specific techniques employed:

> [PAR] places emphasis upon knowledge 'from below,' takes lived experience as the starting point for investigation, values the knowledge produced through collaboration in action, pushes scholarship to be accountable to the communities most affected by it, and may contribute to social change.
>
> (Cahill, 2007, p. 268)

Generally speaking, PAR aims to make research processes more relevant and meaningful. It does this by prioritising the skills, knowledge, and agency of participants. By focusing on collaboration and action, PAR acts "as a vehicle for liberation, radical social transformation and the promotion of solidarity" (Chatterton, Fuller, & Routledge, 2007, p. 218).

ABR – an important subset of PAR – uses diverse expressive modes to encourage participant-centred meaning-making. It represents "an effort to extend beyond the limiting constraints of discursive communication in order to express meanings that otherwise would be ineffable" (Barone & Eisner, 2012, p. 2). ABR's greatest strength is its ability to evoke strong cognitive and emotional responses, not just from workshop participants but also from researchers, collaborators, and public audiences.

Although ABR/PAR makes noble efforts to democratise knowledge production, it should not be considered a miracle salve for the ethical, methodological, and epistemological quandaries plaguing social research. As Lesley-Anne Gallacher and Michael Gallagher (2008) note, participatory epistemologies are grounded in the belief that "people are transparently knowable to themselves" (p. 502) and therefore risk privileging participant voices as the only authentic source of knowledge. There is also hubris in assuming that participants share the same objectives as ABR/PAR researchers, or that they have the time, interest, and resources to be meaningfully involved. For these reasons, Shannon Walsh (2016) regards many ABR/PAR claims, such as the ability to empower oppressed communities or bestow voice on the silenced, as repressive myths.

Although cognisant of these debates, I recognise the enormous potential of creative expression to facilitate learning and promote critical consciousness (see Khan & Marnell, 2022). ABR presents useful opportunities for exploring complex social phenomena in ways that honour the affective, sensory, and imaginative aspects of human experience. As Elsa Oliveira (2016) argues, ABR – when used appropriately and responsibly – can "unveil a wealth of information that other more traditional methods are unable to offer" (p. 276).

Introducing the Project: Arts-Based Research in Action

The reflections that follow emerge from an intensive storytelling workshop with eight LGBTQ+ migrants. It forms part of a larger corpus of work known as the MoVE Project, housed at the African Centre for Migration and Society.[4] The workshop ran for four weeks, during which time participants were introduced to various expressive modes, including narrative writing, applied drama, visual art, and symbolic/spatial mapmaking. To protect privacy, participants were encouraged to use pseudonyms, although some opted to use their real names since this information is already on the public record. Participants were also free to decide whether country-of-origin details or other potentially identifying information would be included in public outputs. Thus, a participant's country of origin is only disclosed here if permission was explicitly granted.

The workshop culminated in the production of public zines. These are low-budget, DIY publications that typically deal with controversial or niche topics (Oliveira & Vearey, 2016). For Barbara Guzzetti and Margaret Gamboa (2004), zines represent "an act of civil disobedience; a tool for inspiring other forms of activism; and a medium through which [their creators] effect changes within themselves" (p. 411). When developing their zines, participants were encouraged to identify a target audience, consider their intended message(s), and experiment with different visual and textual layouts.

Lessons from the Field: Queering Arts-Based Research

Below, I chart five ways that queer theory informs my ABR praxis, drawing on the aforementioned case study. I do not analyse the zines as narrative and visual works as this has been done elsewhere (see Marnell, 2022).

Conceptualisation and Planning

From the outset I saw this project as a political intervention grounded in social activism. This framing encouraged me to develop a workshop process that prioritised critical consciousness over creative or scholarly outputs. Neoliberal academic policies glorify "partnerships," "outcomes," and "impacts" – concepts interpreted in the most obvious and rigid of ways – but I was interested in how this workshop might *evolve* rather than in what it might *produce*. Queering the project meant making space for different needs and interests, even if participants' desires sometimes clashed with my personal vision.

In practical terms, this meant findings ways to incorporate popular education strategies alongside more familiar ABR techniques (see Marnell & Khan, 2016). I also wanted to ensure participants were free to explore unexpected avenues of inquiry. Thus, rather than enter the workshop with a prescriptive schedule and fixed objectives, I sketched out a rough plan of activities that could be adapted as required. This left room for the "complex, diffuse and messy" (Law, 2004, p. 2) aspects of social research.

Initial sessions were dedicated to building a collective analysis of inequality. We began by interrogating heteronormativity in everyday life and then expanded our analysis to other forms of social regulation. This heuristic framework allowed participants to unpack interlinked systems of oppression. It also served as a queer political foundation, in that participants were thinking about the production, operation, and circulation of norms from the outset. This combination of creative and pedagogical activities influenced participants' approach to storytelling: they not only reflected on the power of stories to reinforce or challenge inequality but also considered the social scripts that shape our identities, values, desires, and relationships.

The remainder of the workshop was left open-ended so that participants could apply their knowledge and skills in creative ways. I continued to play an active role by selecting and facilitating activities based on participant feedback, but I tried as much as possible to let the workshop develop in a nonlinear trajectory. Sometimes this meant coming up with activities on the spot or finding ways to balance competing priorities. Put another way, I tried not to fixate on tangible outcomes, but rather on how the workshop process was being negotiated and modified.

On Intimacy and Recognition

Conventional research is grounded in one of the most pernicious conceits of the Enlightenment: that of the "individuated liberal humanist research subject" (Butz, 2008, p. 239). This idea continues to dictate the ethical norms of empiricism, serving as the basis for misguided claims to "objectivity" and "truth." Yet, as Oliveira (2019) rightly points out, what we think and do as researchers is never neutral: "what we might consider 'common sense' is actually informed by a range of cultural backgrounds, power structures and judiciary realities" (p. 527). As a queer researcher, I often struggle to reconcile my ethical obligations as set out by university administrations with my personal politics. I remain suspicious of the researcher–researched dichotomy on which ethics frameworks are based, and I reject the idea that I can somehow extricate myself from hegemonic social structures. Such concerns will be familiar to anyone acquainted with feminist, decolonial, and participatory methodologies. As noted above, ABR/PAR critiques the notion of a detached researcher by adopting "a collaborative and nonhierarchical approach" (Pain, 2004, p. 652) in which ownership of knowledge is negotiated and shared. Participatory approaches also call on researchers to recognise their positionality and to act with critical self-reflexivity (Cahill, Sultana, & Pain, 2007). On one level, then, my ethical concerns are addressed by ABR/PAR, in that my approach is based on reciprocal relationships that prioritise the skills, knowledge, and desires of all parties. Yet, ABR/PAR falls short of what I need as a researcher working simultaneously within and outside of my own community.

Queer theorists offer productive readings of intimacy within research engagements (e.g., Povinelli, 2006). They note that nuanced interpersonal relations are a precursor to situated, entangled, and collaborative forms of knowledge:

> A 'politics of intimacy' as an outcome and progenitor of research becomes a queer project in its disruption of normative considerations for research relationships, while challenging the conventional regimes of oversight for research methods.
>
> (Detamore, 2016, p. 173)

For me, intimacy is both a political and ethical imperative. It extends beyond hollow references to "positionality" and "collaboration" and instead tethers researchers to participants in ways that produce new relational terrains. Troubling the objectivity demanded by conventional methodologies can result in strong emotional bonds that in some instances resemble kinship. These attachments emerge from points of shared identification and take shape through an evolving sense of recognition, connection, and obligation.

More than anything, intimacy requires openness and vulnerability. It is about seeing and being seen, witnessing and experiencing, and ultimately allowing oneself to be affected (Favret-Saada, 2012). Such moments are only possible if one makes a choice to find oneself within others and to push back against research norms that partition affect from knowledge. Oliveira (2019) calls this a "love ethic" and notes how such a thing is only possible when we operate between the arbitrary lines dividing researchers from participants, institutions from communities, theory from praxis. The intimate bonds forged in these in-between spaces can produce new sensibilities, transforming research encounters into sites of radical connection.

So, what does queer intimacy look like in practice? From the beginning of this project, I cultivated moments of recognition by disclosing my identity and experiences in similar ways to participants. It was important for me to take part in each and every storytelling process; to share my own memories, emotions, and aspirations; and to open my reflections to collective analysis. I also made time to discuss my academic and political work; to clarify my motivations for running the project; and to recount the social, educational, and professional trajectories that had brought me to this point in my life.

Moments of connection soon began to occur, taking the form of impromptu dances and songs; playful conversations about sex, relationships, and parties; late-night WhatsApp debates about religion, politics, and law; and break times peppered by spontaneous eruptions of laughter. There were also times spent working through tough emotions like anger and frustration, as well as more joyous occasions when we revelled in each other's personal and artistic growth.

While strong emotional bonds are not uncommon within ABR/PAR, they often take distinct forms within queer spaces. In this project, there were moments of intense recognition. Shared knowledge of what it is like to be a queer body in a heteronormative world seemingly collapsed the distance between researcher and researched, subject and object. When one participant spoke of homophobic violence motivated by his effete gestures, I shared moments in which my own mannerisms have drawn ire and abuse. Others joined in and soon began modelling camp/butch movements on an improvised catwalk. This gleeful, unrehearsed moment not only released the tension but also strengthened the bonds that had started to form. In that moment, our personal differences seemed to evaporate, leaving behind a sense of shared queerness.

Cultivating intimacy is incredibly important to me as a researcher, but it should not be seen as a substitute for serious interrogations of power and privilege (see Kesby, Kindon, & Pain, 2007). Familiar experiences can foster a congenial and productive environment, but they do not alleviate structural disparities. As a white Australian cisgender man, I possess economic and social capital that participants can only imagine – a reality that cannot be ignored. It would also be misleading to suggest that the bonds developed during a workshop immediately translate into friendships. A blurring of hierarchies within research can create beautiful connections, but it can also bring tension and the potential for exploitation. While the participants and I shared many commonalities, our experiences as racialised bodies within colonial histories shaped our interactions. As Ahmed (2000) notes, "friendships and alliances will always take place in situations of asymmetry of power" (p. 58). No matter how progressive my queer politics are, or how well intentioned a project is, or how benevolent I may act, my whiteness invariably creates work for those around me (see de Leeuw, Cameron, & Greenwood, 2012).

Moments of intense recognition can also make it hard for a researcher to set boundaries, especially when working with individuals who face poverty and marginalisation. Yet, rather than distance itself from such complexities, a queer-participatory approach demands discussion, refection, and action. When researching gender and sexuality there is a temptation to minimise other power differentials, such as race and class, but queering research means recognising and responding to interactions between these social locations. To suggest that shared queerness is enough to overcome asymmetries of power is to flagrantly disregard intersectional realities. The same applies in reverse: while ABR/PAR practitioners are often attuned to the nuances of race and class, many seem uninterested in, or at best ambivalent towards, heteronormative social entanglements.

Queering the Storytelling Process

In addition to generating rich data, I hoped that any creative artefacts would spotlight participants' struggles, achievements, and dreams. A dearth of accurate narratives is one of the primary reasons that anti-LGBTQ+ myths persist on the African continent. In that sense, I was inspired by Chinua Achebe's (2000) call for a "balance of stories" through which oppressed people can "contribute to a definition of themselves" instead of being "victims of other people's accounts" (p. 2). Yet, I was also aware of the immense pressure that LGBTQ+ migrants face when narrating their experiences. To be eligible for state protection, they must find ways to make their identities and histories legible for bureaucrats, even when hypervisibility brings violence and exploitation (Marnell, 2023; Camminga, 2019).

Conscious of these complexities, I looked for ways to queer the storytelling process. The goal was not only to collect diverse narratives but also to experiment with representational forms that challenge mainstream discourses. This was achieved by incorporating diverse expressive modes, each of which offered a new avenue for depicting and analysing participants' social worlds. The resulting narratives questioned the single story of LGBTQ+ migration (Marnell, 2021). The participants and I probed both the mundane and the fabulous aspects of our lives, while also making space for affective, sensual, and quixotic dreamscapes. This is not to say that I pushed participants to produce "queer" narratives. Rather, it points to the radical imaginings that can erupt out of queer intimacy. Iterative processes of creating, sharing, analysing, and revising – run in parallel with the pedagogical activities described above – produced unexpected, deeply poignant narratives. Some participants explored topics/themes not usually associated with queer migration (e.g., cooking and clothing), while others opted to share transgressive or controversial anecdotes (e.g., sexual trysts and drug taking).

The decision to make public zines reflected participants' growing interest in using storytelling to challenge social conventions. During the early stages of the project, I suggested the group develop a collective publication, possibly a guidebook for newly arrived LGBTQ+ migrants. However, the group rejected this idea, opting to create individual publications based on their personal experiences. The visual language employed by participants (distorted graphics, cut-out letters, and hand drawings), combined with a largely nonlinear presentation of written narratives, created a distinctly queer aesthetic. The zines embraced pastiche and blending, reflecting the disorderliness of life and the fluidity of identities; they are sorrowful and despairing, yet they are also optimistic, joyful, and sensual.

A notable example is *The Trials of Tino*, a zine by a gay asylum seeker from Zimbabwe. On one double-page spread, Tino juxtaposes experiences of violence with a detailed account of an orgy, complete with playfully repurposed magazine headlines (Figures 4.1 and 4.2). He also interspersed his zine with collaged pictures of himself in various poses, representing the gamut of emotions he experiences as a gay asylum seeker facing both homophobia and xenophobia. A similar disruption of mainstream narratives occurs in Dee Jay's zine, *Life Is Not About Waiting*. Dee Jay, a lesbian asylum seeker, explores the spatial elements of identity, showing how her gender and sexuality are experienced in different locations – at home, in church, on the street. By highlighting different facets of their lives, Tino and Dee Jay mess with hegemonic identity configurations and challenge stereotypes linking LGBTQ+ identities with sickness, depravity, promiscuity, and blasphemy.

By the end of the workshop, participants not only felt comfortable producing stories about their lives but also wanted to inspire other LGBTQ+ migrants to do the same. In a follow-up interview, Hotstix, a lesbian woman from Zimbabwe (Figures 4.3 and 4.4), affirmed her belief in the transformative power of storytelling:

> It's important for us to tell more stories because if we don't share about our lives, people in the community who don't know about us will never have an idea of what we go through. I actually urge every LGBTQ+ person to tell about what they go through in their day-to-day lives. I want my brothers and sisters to feel strong enough to share all the stuff – the good, the bad, and the ugly. We shouldn't hide who we are.

Of note here is Hotstix's assertion that LGBTQ+ persons must push back against dominant scripts by celebrating all aspects of life: the good, the bad, and the ugly.

Thinking, Doing, and Feeling

Like Muñoz (1996) and Ahmed (2006), I believe a queer praxis requires a distinct orientation. This encompasses both alignment (our position in relation to people, texts, objects, emotions, and behaviours) and sensibility (our ability to recognise, appreciate,

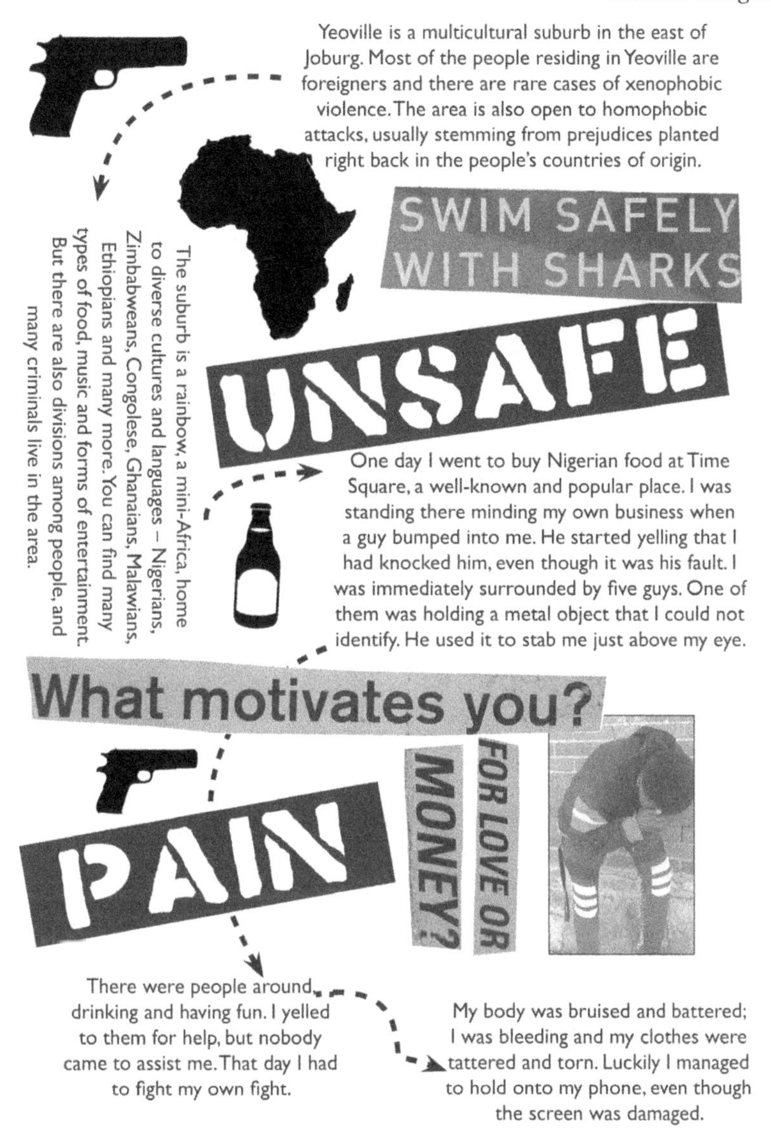

Figure 4.1 Left side of zine spread about Tino's life in Yeoville, Johannesburg

and respond to particular cues, be they verbal or nonverbal). A queer orientation does not come from institutional research training or methodological textbooks, but rather emerges from a political standpoint. It refers not just to how we perceive things but also how we negotiate relational trends. A large part is reflecting honestly on the direction from which we enter research. As Alison Rooke (2016) notes, "queer reflexivity requires drawing attention to the erotics of knowledge production" (p. 35). This means recognising the bodies, affects, and politics circulating within and around a research process. A queer orientation is sensitive to how heteronormative discourses, relations of power, and vocabularies of meaning not only shape participation but also influence the production and interpretation of knowledge.

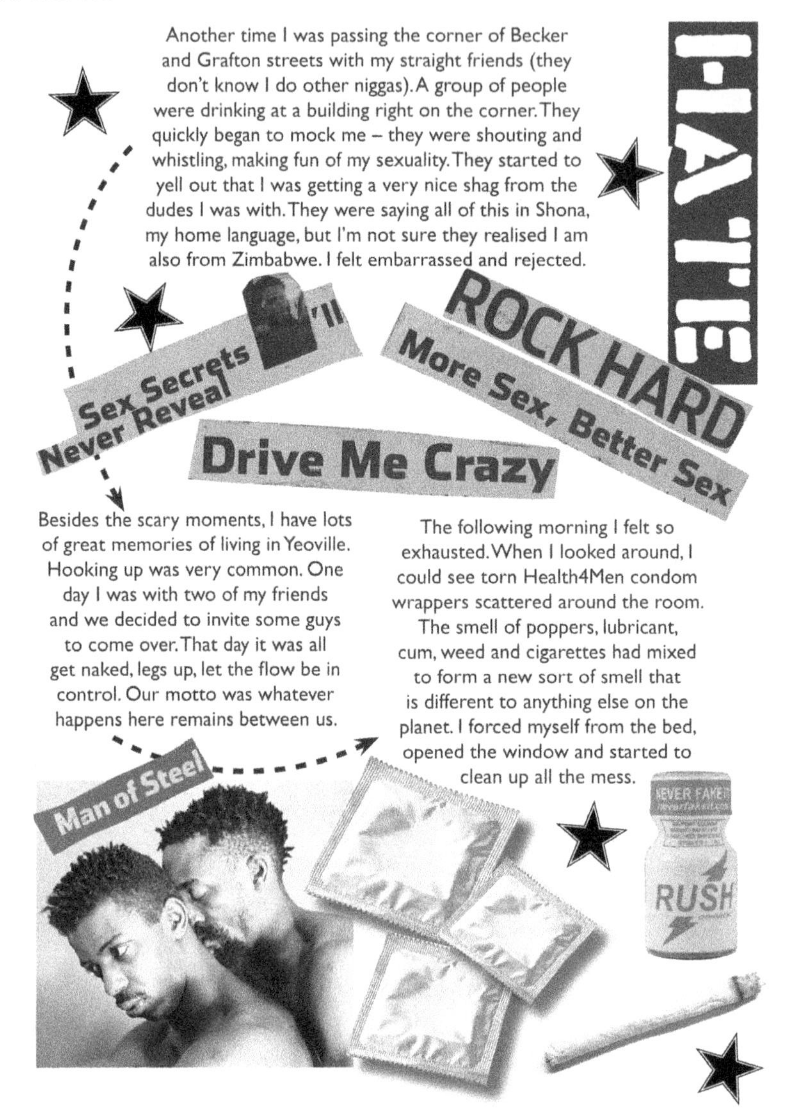

Figure 4.2 Right side of zine spread about Tino's life in Yeoville, Johannesburg

As a queer researcher, I consider myself attuned to the nuances of gender and sexuality, yet this is not enough to navigate the intricacies of every encounter, especially those involving multiple sociocultural borders. Throughout the workshop I was forced to re-examine my politics and perspectives; like everyone, I am susceptible to the ontological comfort that comes with seeing categories as fixed and coherent. Thus, it was important for me to consciously reflect on the performativity of my own subject position.

Describing a queer sensibility is not easy. Often it is a matter of perceiving and processing a glance, a movement, or a choice of words. It is about not taking at face value what is seen, heard, or felt, but instead recognising that normative significations can carry hidden meanings. At other times, it is about confronting the reality that others may

Figure 4.3 Inside front cover of Hotstix's zine, titled *Survival of the Fittest*

construct and express their identities in ways that jar with more familiar configurations. When Mike, a gay asylum seeker, spoke of his Christian faith, I had to find ways to open religious discourses to collective analysis without devaluing this aspect of his selfhood. Similarly, when Hotstix brought a new girlfriend to the workshop, I chose to read her actions as productive rather than disruptive. Observing Hotstix's behaviours, I realised she wanted to affirm both her butch identity and her status within the group.

My responses to these moments were driven by my queer politics, but they were also mediated through my social position, personal feelings, embodied sensations, and cognitive experiences (Ahlstedt, 2015). There were likely times when I misread or disregarded participants' needs. This is one of the challenges of being situated both within and outside of a community. My status as a queer migrant brought me closer to participants, but my race, class, and nationality ensured a gulf in our experiences. In other words, my

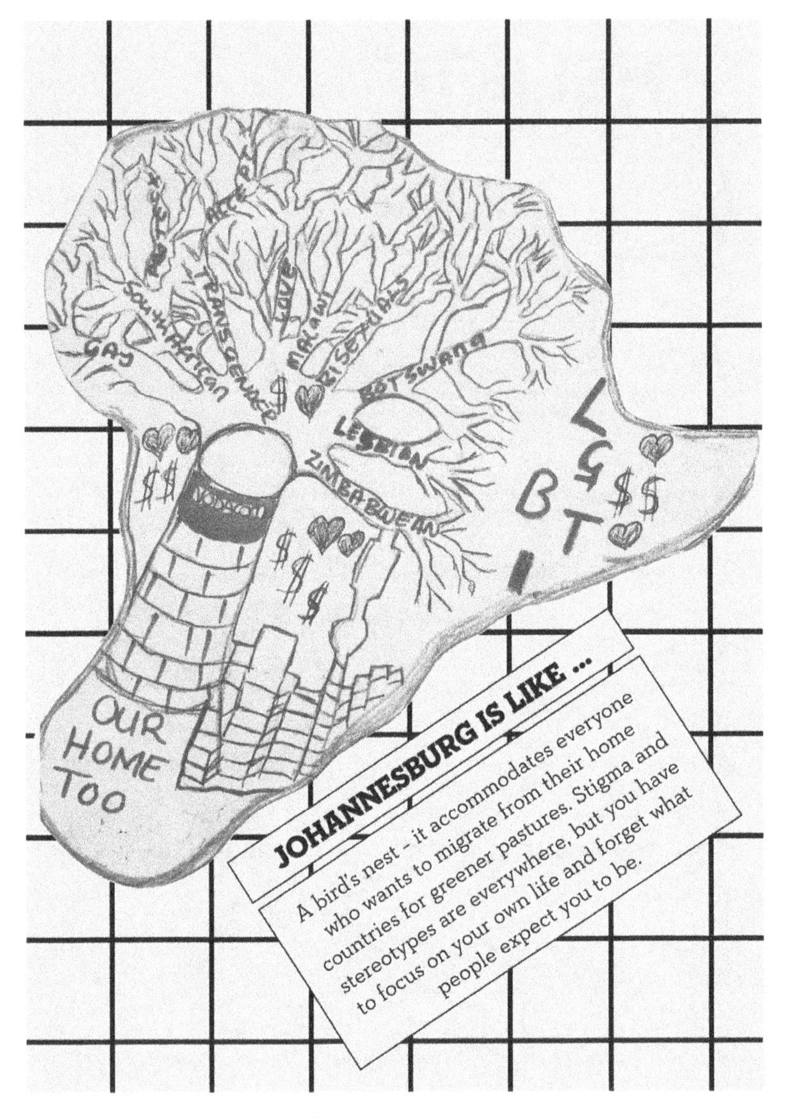

Figure 4.4 Back cover of Hotstix's zine, featuring a hand-drawn image

queer orientation made certain utterances, gestures, and interactions decipherable while others remained illegible and perhaps invisible.

A queer orientation should also extend beyond the physical confines of a workshop space, guiding processes such as the presentation and dissemination of findings. Instead of "writing up" my research, I have experimented with different ways of "writing in" (Mansvelt & Berg, 2010). Like many ABR/PAR practitioners, I argue for participants to be meaningfully included in the analysis of creative works and for their voices to be at the centre of project outputs (Marnell, Oliveira, & Khan, 2021). But a queer-participatory methodology requires more than this if it is to unsettle hegemonic epistemologies. In both my general and academic writing, I reject the scholarly convention of using the authoritative third person. Rather, I try to position myself as a gendered, sexed, and raced body moving in and around research processes. Queering academic writing means foregrounding the political and ethical dimensions of our praxis and acknowledging the emotional intimacies that drive such engagements.

Practically speaking, I resist the temptation to frame workshops as straightforward and triumphant, or to portray participants' subjectivities as fixed and stable. For example, Jonso's zine and follow-up interview include self-identifications as both a lesbian woman and a transgender man (Figures 4.5 and 4.6). Rather than present these categories as antithetical, I allow them to sit side by side, reflecting as they do a complex process of being and becoming. Similarly, when describing Hotstix, I make space for the smiles, laughs, dances, jokes, frustrations, ambitions, loyalty, and swagger I witnessed during the workshop. Rich descriptions such as these are vital if I am to honour the multifaceted people I have come to know and admire. This style of writing not only pushes back against academic norms but also complicates assumptions about African LGBTQ+ migrant lives.

Queering Notions of Change

While I hope my research contributes to positive social change for LGBTQ+ persons (as a feminist or intersectional approach may aim to improve material conditions for women or people of colour), I understand the transformative potential of my work differently from many ABR/PAR practitioners. First and foremost, I am interested in how the identities at the heart of my scholarly endeavours – "gay" or "trans" or "woman" or "migrant" or "refugee" – are discursively constructed, as evidenced by their inconsistencies and incoherencies. Orienting my work in this way may put me at odds with classical sociology, most notably in my distrust for orthodox readings of subjectivity, but it also opens up opportunities to explore more subtle modes of resistance.

A queer-participatory approach can also generate new ways of conceptualising power dynamics. Confronting inequalities is a cornerstone of ABR/PAR, as Andrea Cornwall and Rachel Jewkes (1995) note: "participatory research is primarily differentiated from conventional research in the alignment of power" (p. 1668). However, much of the literature offers impoverished theorisations of power. Queer theory's focus on regulatory regimes serves as a useful corrective, showing that power operates through circuits of meaning rather than as a fixed capacity. Of particular importance is Michel Foucault's vision of power as "a diverse, ambivalent web of relations, rather than a unidirectional force of domination" (Gallagher, 2008, p. 144). Such a reading reminds us that "we can never be ensnared by power: we can always modify its grip" (Foucault, 1988, p. 123).

I found this perspective useful when thinking about relational dynamics within the project, not only between participants and me but also among participants themselves. Such tensions were evident when the group was deciding how to disseminate their zines. A vigorous debate took place about the political imperative to take up space and demand recognition vis-à-vis the more palatable "we-are-just-like-you" approach. I was struck by the how participants formed and argued different positions, drawing on earlier group discussions. For example, Tino advocated for confrontational forms of activism. He was determined to keep the more explicit stories in his zine and to leave the finished product in random places. Others felt this would alienate potential readers and further entrench social divides. The waves of tension

Another time I was struggling to find accommodation. I didn't have any money because I was not working, and I had to sleep at friends' places in Hillbrow, Berea and Soweto. I spent four months going from friend to friend. I was embarrassed, but I didn't have a choice. Staying at other people's houses is hard: you have to follow their rules and you don't have any privacy. Imagine what it's like when it's raining – you have to wait for the owner to come back from work and open up for you. It's hurtful and humiliating.

HARD TIMES

NEVER KILL YOU

BUT MAKE YOU STRONGER

Figure 4.5 Zine extract about Jonso's housing challenges in Johannesburg

and resolution I observed during this debate highlighted just how nuanced power relations are within a creative research environment. I encouraged participants to weigh up different forms of resistance, and it soon became clear that radical disruptions (such as forcing queer representations into the public realm) can be as important to participants as traditional advocacy.

Finding ways to link research to social transformation is not new for ABR/PAR practitioners, but a queer-participatory approach may offer something unique in how we conceptualise action. In many projects, change is reduced to improving access to services or reforming state institutions. By contrast, a queer framing invites participants to probe and resist established categories. In saying this, I do not wish to devalue economic development or material change. Rather, I seek to highlight how a queer-participatory approach can unlock new forms and sites of activism.

Figure 4.6 Inside back cover of Jonso's zine, titled *I Never Give Up*

Conclusion: A Queer Vision for ABR/PAR

In reflecting on this project, I hope to have demonstrated the value of queering ABR/PAR. While participatory approaches represent a challenge to dominant empirical and methodological paradigms, their heteronormative underpinnings remain largely unchecked. A queer-participatory framing can amplify existing critiques of positivism while simultaneously destabilising hegemonic assumptions and established practices. In particular, a queer-participatory approach can unsettle rigid definitions of what it means to *do* research and foster opportunities for new types of collaboration and creativity. By valuing the affective and corporal dimensions of human experience, by questioning assumptions about subjectivity and identity, and by interrogating the impact of regulatory regimes on knowledge production, a queer orientation can transform how and why we do ABR/PAR. These possibilities exist regardless of the topic under investigation or the individuals/communities involved. Most importantly, a queer praxis reminds us to never blindly accept business as usual within the academy, even when that business claims to be participatory.

Notes

1. I would like to sincerely thank the rights holders for granting permission to publish the two epigraphs. For the original source material, see bell hooks (2006) and Giuseppe Campuzano and Miguel Á. López (2014). I strongly encourage readers to view the artwork from which the second epigraph has been quoted (see https://www.museoreinasofia.es/en/collection/artwork/beauty-salon).
2. "LGBTQ+ migrant" is used as an umbrella term for "lesbian, gay, bisexual, transgender, and queer migrants, refugees, and asylum seekers." This is done in recognition that established legal categories do not necessarily align with lived experiences or reflect individuals' self-identifications. In contexts like South Africa, bureaucratic failings push potential asylum seekers into the immigration system and vice versa. The "correct" terms are used when clarification is needed or when referencing specific individuals.
3. Scholars from many socio-cultural backgrounds have critiqued and reconfigured queer theory. I foreground work from the African continent because of the location of this project. I do so without wishing to disregard the contributions of queer Indigenous and two-spirit researchers, activists, and communities.
4. MoVE experiments with different visual and narrative methods to research the lived experiences of migrants (see www.facebook.com/themoveprojectsouthafrica).

References

Achebe, C. (2000). *Home and Exile*. Oxford: Oxford University Press.

Acosta, K. L. (2018). Queering family scholarship: Theorizing from the borderlands. *Journal of Family Theory & Review*, 10(2): 406–418.

Ahlstedt, S. (2015). Doing 'feelwork': Reflections on whiteness and methodological challenges in research on queer partner migration. In R. Andreassen & K. Vitus (Eds.), *Affectivity and Race: Studies from a Nordic Context*. 187–203. Farnham: Ashgate.

Ahmed, S. (2000). Who knows? Knowing strangers and strangeness. *Australian Feminist Studies*, 15 (31): 49–68.

Ahmed, S. (2006). *Queer Phenomenology: Orientations, Objects, Others*. Durham, NC: Duke University Press.

Barone, T., & Eisner, E. W. (2012). What is and what is not arts-based research? In T. Barone & E. W. Eisner (Eds.), *Arts-Based Research*. 1–12. Thousand Oaks, CA: Sage.

Boyce, P., Engrebretsen, E. L., & Posocco, S. (2018). Introduction: Anthropology's queer sensibility. *Sexualities*, 21(5/6): 843–852.

Brim, M., & Ghaziani, A. (2016). Introduction: Queer methods. *Women's Studies Quarterly*, 44(3/4): 14–27.

Browne, K., & Nash, C. J. (2016). Queer methods and methodologies: An introduction. In K. Browne & C. J. Nash (Eds.), *Queer Methods and Methodologies: Intersecting Queer Theories and Social Science Research*. 1–24. London: Routledge.

Butler, J. (1990). *Gender Trouble: Feminism and the Subversion of Identity*. New York: Routledge.

Butler, J. (1993). *Bodies that Matter: On the Discursive Limits of "Sex."* New York: Routledge.

Butz, D. (2008). Sidelined by the guidelines: Reflections on the limitations of standard informed consent procedures for the conduct of ethical research. *ACME*, 7(2): 239–259.

Cahill, C. (2007). The personal is political: Developing new subjectivities through participatory action research. *Gender, Place & Culture*, 14(3): 267–292.

Cahill, C., Sultana, F., & Pain, R. (2007). Participatory ethics: Politics, practices, institutions. *ACME*, 6(3): 304–318.

Campuzano, G., & López, M.Á. (2014). *Beauty Salon*. Madrid: Collection Museo Nacional Centro de Arte Reina Sofía.

Camminga, B. (2019). *Transgender Refugees and the Imagined South Africa: Bodies over Borders and Borders over Bodies*. London: Palgrave Macmillan.

Chatterton, P., Fuller, D., & Routledge, P. (2007). Relating action to activism: Theoretical and methodological reflections. In S. Kindon, R. Pain, & M. Kesby (Eds.), *Participatory Action Research Approaches and Methods: Connecting People, Participation and Place*. 216–222. New York: Routledge.

Cornwall, A., & Jewkes, R. (1995). What is participatory research? *Social Science & Medicine*, 41 (12): 1667–1676.

Crosby, C., Duggan, L., Ferguson, R., Floyd, K., Joseph, M., Love, H., *et al.* (2011). Queer studies, materialism, and crisis: A roundtable discussion. *GLQ: A Journal of Lesbian and Gay Studies*, 18 (1): 127–147.

Dadas, C. (2016). Messy methods: Queer methodological approaches to researching social media. *Computers and Composition*, 40: 60–72.

Dahl, U. (2016). Femme on femme: Reflections on collaborative methods and queer femme-inist ethnography. In K. Browne & C.J. Nash (Eds.), *Queer Methods and Methodologies: Intersecting Queer Theories and Social Science Research*. 143–166. Oxon: Routledge.

de Leeuw, S., Cameron, E. S., & Greenwood, M. L. (2012). Participatory, community-based research, indigenous geographies, and the spaces of friendship: Sites of critical engagement. *The Canadian Geographer*, 56(2): 180–194.

Detamore, M. (2016). Queer(y)ing the ethics of research methods: Toward a politics of intimacy in researcher/researched relations. In K. Browne & C. J. Nash (Eds.), *Queer Methods and Methodologies: Intersecting Queer Theories and Social Science Research*. 167–182. Oxon: Routledge.

Di Feliciantonio, C., Gadelha, K. B., & DasGupta, D. (2017). Queer(y)ing methodologies: Doing fieldwork and becoming queer. *Gender, Place & Culture*, 24(3): 403–412.

Ekine, S., & Abbas, H. (2013). Introduction. In S. Ekine & H. Abbas (Eds.), *Queer African Reader*. 1–5. Dakar: Pambazuka.

Favret-Saada, J. (2012). Being affected. *HAU*, 2(1): 435–445.

Fields, J. (2016). The racialized erotics of participatory research: A queer feminist understanding. *Women's Studies Quarterly*, 44(3/4): 31–50.

Fish, J. N., & Russell, S. T. (2018). Queering methodologies to understand queer families. *Family Relations*, 67(1): 12–25.

Foster, V. (2016). *Collaborative Arts-Based Research for Social Justice*. Oxon: Routledge.

Foucault, M. (1988). Power and sex. In L. Kritzman (Ed.), *Michel Foucault: Politics, Philosophy, Culture, Interviews and Other Writings, 1977–1984*. 110–124. London: Routledge.

Frisby, W., Maguire, P., & Reid, C. (2009). The 'f' word has everything to do with it: How feminist theories inform action research. *Action Research*, 7(1): 13–29.

Gallacher, L., & Gallagher, M. (2008). Methodological immaturity in childhood research? Thinking through 'participatory methods'. *Childhood*, 15(4): 499–516.

Gallagher, M. (2008). 'Power is not an evil': Rethinking power in participatory methods. *Children's Geographies*, 6(2): 137–150.

Gamson, J. (2000). Sexualities, queer theory, and qualitative research. In N. K. Denzin & Y. S. Lincoln (Eds.), *Handbook of Qualitative Research*. 347–365. Thousand Oaks, CA: Sage.

Giametta, C. (2018). Reorienting participation, distance and positionality: Ethnographic encounters with gender and sexual minority migrants. *Sexualities*, 21(5/6): 868–882.

Green, A. (2002). Gay but not queer: Toward a post-queer study of sexuality. *Theory & Society*, 31 (4): 521–545.

Guzzetti, B. J., & Gamboa, M. (2004). Zines for social justice: Adolescent girls writing on their own. *Reading Research Quarterly*, 39(4): 408–436.

Halberstam, J. (1998). *Female Masculinity*. Durham, NC: Duke University Press.

Halberstam, J. (2011). *The Queer Art of Failure*. Durham, NC: Duke University Press.

hooks, b. (2006). *Outlaw Culture: Resisting Representations*. New York: Routledge.

Johnson, E. P. (2016). Put a little honey in my sweet tea: Oral history as quare performance. *Women's Studies Quarterly*, 44(3/4): 51–67.

Kesby, M., Kindon, S., & Pain, R. (2007). Participation as a form of power: Retheorising empowerment and spatialising participatory action research. In S. Kindon, R. Pain, & M. Kesby (Eds.), *Participatory Action Research Approaches and Methods: Connecting People, Participation and Place*. 19–25. New York: Routledge.

Khan, G. H., & Marnell, J. (2022) Reimagining wellbeing: Using arts-based methods to address sexual, gender and health inequalities. *Global Public Health*, 17(10): 2574–2589.

Law, J. (2004). *After Method: Mess in Social Science Research*. London: Routledge.

Lenette, C. (2019). *Arts-Based Methods in Refugee Research: Creating Sanctuary*. Singapore: Springer.

Macharia, K. (2015). Archive and method in queer African studies. *Agenda*, 29(1): 140–146.

Mansvelt, J., & Berg, L. D. (2010). Writing qualitative geographies, constructing meaningful geographical knowledges. In I. Hay (Ed.), *Qualitative Research Methods in Human Geography*. 333–355. Oxford: Oxford University Press.

Marnell, J. (2023). City streets and disco beats: Recentring the urban in queer and trans migration studies. *Urban Forum*, 23: 201–211.

Marnell, J. (2022). Telling a different story: On the politics of representing African LGBT+ migrants, refugees and asylum seekers. In B. Camminga & J. Marnell (Eds.), *Queer and Trans African Mobilities: Migration, Asylum and Diaspora*. 39–60. London: ZED Books.

Marnell, J. (2021). *Seeking Sanctuary: Stories of Sexuality, Faith and Migration*. Johannesburg: Wits University Press.

Marnell, J., & Khan, G. H. (2016). *Creative Resistance: Participatory Methods for Engaging Queer Youth*. Johannesburg: GALA.

Marnell, J., Oliveira, E., & Khan, G. H. (2021). 'It's about being safe and free to be who you are': Exploring the lived experiences of queer migrants, refugees and asylum seekers in South Africa. *Sexualities*, 24(1/2): 86–110.

Matebeni, Z., Monro, S., & Reddy, V. (Eds.) (2018). *Queer in Africa: LGBTI Identities, Citizenship and Activism*. New York: Routledge.

McCann, H. (2016). Epistemology of the subject: Queer theory's challenge to feminist sociology. *Women's Studies Quarterly*, 44(3/4): 224–243.

McLean, H. (2018). In praise of chaotic research pathways: A feminist response to planetary urbanization. *Environment and Planning D: Society and Space*, 36(3): 547–555.

McWilliams, J. (2016). Queering participatory design research. *Cognition and Instruction*, 34(3): 259–274.

Muñoz, J. E. (1996). Ephemera as evidence: Introductory notes to queer acts. *Women & Performance*, 8(2): 5–16.

Nyanzi, S. (2014). Queering queer Africa. In Z. Matebeni (Ed.), *Reclaiming Afrikan: Queer Perspectives on Sexual and Gender Identities*. 65–68. Athlone: Modjaji Books.

Nyeck, S. N., & Epprecht, M. (2013). Introduction. In S. N. Nyeck & M. Epprecht (Eds.), *Sexual Diversity in Africa: Politics, Theory, and Citizenship*. 4–15. Montreal: McGill Queens University Press.

Oliveira, E. (2016). Empowering, invasive or a little bit of both? A reflection on the use of visual and narrative methods in research with migrant sex workers in South Africa. *Visual Studies*, 31 (3): 260–278.

Oliveira, E. (2019). The personal is political: A feminist reflection on a journey into participatory arts-based research with sex worker migrants in South Africa. *Gender & Development*, 27(3): 523–540.

Oliveira, E., & Vearey, J. (Eds.) (2016). *The Sex Worker Zine Project*. Johannesburg: The MoVE Project.

Pain, R. (2004). Social geography: Participatory research. *Progress in Human Geography*, 28(5): 652–663.

Povinelli, E. (2006). *The Empire of Love: Toward a Theory of Intimacy, Genealogy, and Carnality*. Durham, NC: Duke University Press.

Rooke, A. (2016). Queer in the field: On emotions, temporality and performativity in ethnography. In K. Browne & C. J. Nash (Eds.), *Queer Methods and Methodologies: Intersecting Queer Theories and Social Science Research*. 25–40. Oxon: Routledge.

Singh, A. A., Richmond, K., & Burnes, T. R. (2013). Feminist participatory action research with transgender communities: Fostering the practice of ethical and empowering research designs. *International Journal of Transgenderism*, 14(3): 93–104.

Taylor, Y. (2016). The 'outness' of queer: Class and sexual intersections. In K. Browne & C. J. Nash (Eds.), *Queer Methods and Methodologies: Intersecting Queer Theories and Social Science Research*. 69–84. Oxon: Routledge.

Tellis, A., & Bala, S. (2015). Introduction: The global careers of queerness. In A. Tellis & S. Bala (Eds.), *The Global Trajectories of Queerness: Re-Thinking Same-Sex Politics in the Global South*. 13–27. Leiden: Brill.

Valocchi, S. (2005). Not yet queer enough: The lessons of queer theory for the sociology of gender and sexuality. *Gender & Society*, 19(6): 750–770.

Walsh, S. (2016). Critiquing the politics of participatory video and the dangerous romance of liberalism. *Area*, 48(4): 405–411.

Warner, M. (1993). *Fear of a Queer Planet: Queer Politics and Social Theory*. Minneapolis, MN: University of Minnesota Press.

5 Mapping Our Home

Using Participatory Mapping to Challenge Police Violence in the South Bronx

Brett Stoudt

Introduction

Ignited by police officers murdering George Floyd and Breonna Taylor, and within the context of a global pandemic accompanying an economic and political crisis, an unprecedented outpouring of people took to the streets across the United States and the world in 2020 and again in 2021. It was a critical moment of collective outrage directed towards the unbearable harms policing has perpetrated on communities of colour for generations. Here in New York City (NYC), crowds marched each consecutive day for months to call for a radical reimagining of public safety. They demanded a significant reduction in the New York Police Department's (NYPD) budget in favour of multibillion-dollar reinvestments into public safety that centres community organisations and social services. New Yorkers are not new to this fight; in fact, the current surge of resistance against state sanctioned violence on communities of colour has long roots in slave patrols and Jim Crow South, and recent roots in a decades-long struggle to end abusive NYPD practices, especially their use of "stop-and-frisk," a police-initiated stop that can escalate to a frisk, search and possibly arrest where the officer has reasonable suspicion that the person will commit, is committing, or has committed a crime (Stoudt, Fine, & Fox, 2011).

By 2010, communities of colour across NYC were beginning to organise in resistance to the NYPD's discriminatory use of stop-and-frisk. By 2012, these communities grew into what would become a strong, successful, and sustained movement against police power and abuse (See www.changethenypd.org). In one small South Bronx neighbourhood in 2010, three mothers were outraged and worried by the police harassment their sons were experiencing on a near daily basis. The mothers' local efforts to challenge police practices grew to involve other residents, and ultimately academic allies from the Public Science Project at the City University of New York (Stoudt, 2016; Stoudt et al., 2015; Stoudt et al., 2019; see also www.publicscienceproject.org). Together, they/we would organise a critical participatory action research (PAR) partnership (Torre et al., 2012; Torre et al., 2018) called the Morris Justice Project. Our aims were to produce data useful for igniting a broader community conversation, as well as to contribute to learning, advocacy, and the police reform movement. Participatory mapping emerged as a central tool for the Morris Justice Project to study and resist police control/power as well as to create spatial accountability structures grounded in the South Bronx.

Maps are ubiquitous in our society and come in many forms, from computer-based Geographic Information Systems (GIS) to three-dimensional sculptures, to markings drawn on paper (Cochrane & Corbett, 2020). Nonetheless, they can have profound impact on how people make place and understand their lives (Crampton, 2011). Maps

DOI: 10.4324/9780429400346-5

are not neutral representations of spatial facts, but instead often serve hegemonic interests that produce and preserve oppressive structures (Smith, 1999; Hunt & Stevenson, 2016). Like statistics, they hold power as legitimising practices, presenting geopolitical information in ways that appear precise, objective, natural, and uncontested (McCall, 2006). Maps are used as political and bureaucratic tools, organised by boundaries such as census blocks or precincts, to manage people, produce oversight, and distribute resources (Crampton, 2011).

The power and implication of maps are acutely illustrated in modern police departments, such as the NYPD. The NYPD employs a fully integrated management system known as CompStat ("compare statistics") that uses sophisticated GIS to monitor, respond to, and even predict crimes in real time (Eterno & Silverman, 2012). Here, the full power of the state to surveil, control, and manage people is on display. So too is their power to regulate the political narrative through data, to define what data are warehoused and how the data are classified (Muhammad, 2019). The authority to include and exclude, define and frame creates, for example, analyses focused on index crime rates but no database fully documenting police violence.

Mapping is a knowledge-producing practice that can exploit as well as illuminate and educate. It can serve emancipatory aims to challenge, disrupt, and even transform oppressive structures. It can also initiate paths for the reclamation, revision, and reimagination of space. Participatory mapping offers a framework towards these aims by attending to power and asking critical questions, such as who produced the maps? For whom were they created? What/whose interests are they furthering? What assumptions are built into the map? And how are the maps incomplete? (Maharawal & McElroy, 2018). There are many variations to participatory mapping, including community mapping, countermapping, participatory GIS, and public participation GIS (Cochrane & Corbett, 2020). Though differences exist, each is committed to an iterative, collaborative mapmaking that centres the grounded expertise and interests of those most marginalised, in the pursuit of understanding, empowerment, and transformation (Dalton & Stallmann, 2018; Kidd, 2019; Maharawal & McElroy, 2018).

This chapter demonstrates an innovative approach to participatory mapping that draws upon "stats-n-action" (Stoudt, 2016; Stoudt et al., 2019; Stoudt, 2014), a set of critical and collaborative quantitative commitments to reframe analysis as an interpretive and exploratory approach that resembles and complements qualitative analysis; one that is inductive, iterative, flexible, contextualised, and relies heavily on visual displays of data (Chambers, 2007; Irvine, Miles, & Evans, 1979; Tukey, 1969). As such, this participatory quantitative mapping process was designed to enact a praxis grounded in the liberatory possibilities of localised numbers, constructed from public and private data, to generate a reflexive dialogue – to learn, theorise, and act (Glass & Stoudt, 2019). The ongoing spatial lens offered through this approach to participatory mapping (re)framed and exposed the contested political terrain of state-sanctioned violence, while shaping physical, relational, ethical, and symbolic boundaries deeply rooted in community.

The carceral state, and police as the primary entry point into the criminal legal system, has contributed heavily to the (re)production of structural oppression against communities of colour in the Unites States (Ritchie, 2017; Vitale, 2017). The spatial analyses of crime and disorder have long been used by police as surveillance mapping technologies

(Desrosières, 1998; Zuberi & Bonilla-Silva, 2008). To counter this, the Morris Justice Project used participatory mapping as a political act of sousveillance (Bradshaw, 2013), taking methodological control of the process to reframe the gaze onto policing as an institution that frequently enacts harmful policies and practices in their neighbourhood. We pursued "critical geographies of home" (Brickell, 2012), holding ourselves spatially accountable to the neighbourhood most of our members called *home* by weaving mapping activities throughout our project.

The Morris Justice Project lasted over five years and was deeply involved in the citywide police reform movement during that time. I use it as a case study for this chapter – inspired by the spatial analyses of Du Bois (2007) and Bunge (2011) – to reflect on participatory mapping as an iterative strategy to learn intimately about *home* and challenge the acute violence of the carceral state. In the first section, I discuss our spatial analyses of the NYPD's stop-and-frisk data. I then review our spatial analyses grounded in community conversations. I conclude by suggesting that participatory action research requires critical accountability structures to enact its critical commitments. For the Morris Justice Project, participatory mapping served to hold us spatially accountable to the local neighbourhood.

Mapping Our Home

Mapping for the Morris Justice Project was a critical component to establishing our participatory action research collective. In the beginning, we met each week in the local public library basement to hold what we called a "research camp." This was concentrated time and space, early in our project, to learn about the South Bronx and begin exploring the local impact of current NYPD policies and practices. We started by mapping NYC policing data over time, revealing a steep increase in low-level police contact predominantly on people of colour. By the time we started regularly gathering in 2011, over 4.4 million summons, 3.5 million stops, and 1.9 million misdemeanours had amassed in just the previous eight years. In the year prior, 601,285 stops were recorded. By the end of our first year together, the NYPD would reach its historic apex (685,724 stops), of which 84% were Black or Latinx, 88% involved neither an arrest nor a summons, and those that did involve the courts were frequently dismissed (Geller, 2015; Schneiderman, NYS Attorney General, & Civil Rights Bureau, 2013).

We next found it informative to analyse police activity spatially, both within and across NYC precincts. Most members of the Morris Justice Project lived in the 44th precinct of the South Bronx, with boundaries that notoriously involved the home of hip hop and Yankees baseball, as well as extreme poverty and urban divestment. It came as no surprise to the group that the 44th precinct was among the most heavily policed in NYC, ranking seventeenth in police stops, fifth in summons, and second in misdemeanour arrests between 2003 and 2010. New Yorkers were repeatedly told by the NYC mayor and police commissioner that stop-and-frisks were indispensable to getting guns off the streets and keeping communities of colour safe (Coates, 2013). Yet our maps told a different story about the 44th precinct, revealing that guns were found in less than 0.2% of the stops. In fact, far from providing safety, we learned that 50% of the police stops in this precinct involved some form of physical force:

(a)

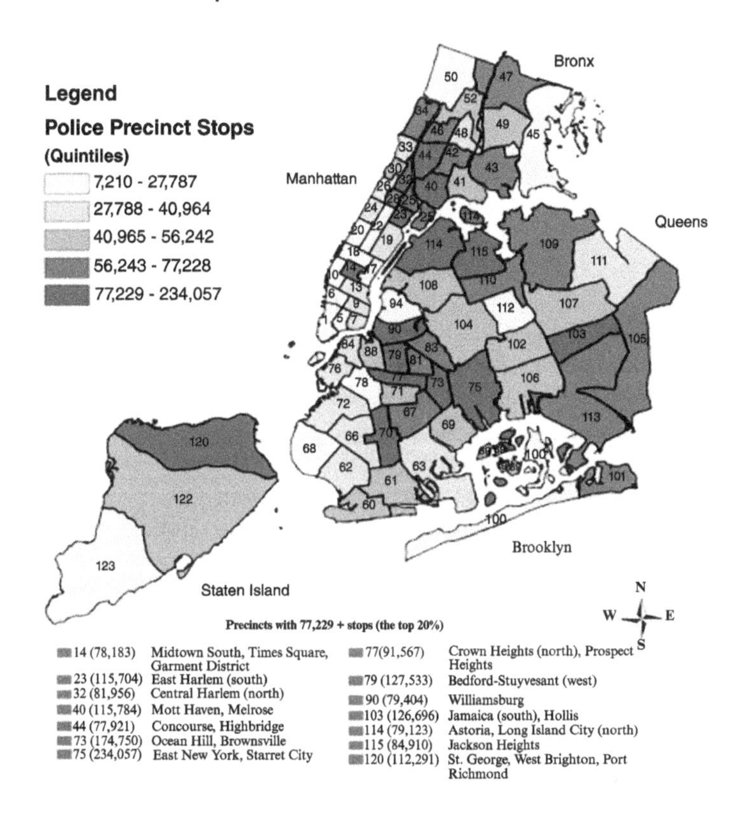

(b)

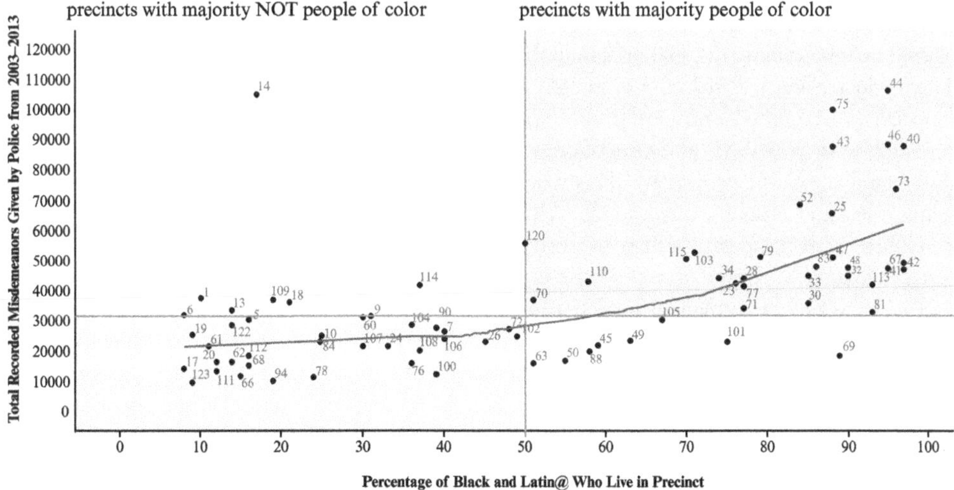

Figure 5.1 Top: accumulated stops over eight years across each precinct color-coded by quintiles. Bottom: accumulated misdemeanours over ten years across each precinct on the y-axis, the x-axis displays the proportion of Black and Latinx residents in the precinct, the star is the 44th precinct

The initial maps we produced helped illustrate the enactment of specific policies designed to surveil and control this community and others like it across the city. They were living the spatially punitive vision of former Mayor Giuliani and Commissioner Bratton's "Police Strategy No. 5: Reclaiming the Public Spaces of New York City" (Giuliani & Bratton, 1994), what would become known as "quality-of-life" policing oriented by Broken Windows Theory (also known as "order-maintenance" or "zero tolerance" policing). The claim was that Broken Windows policing substantially lowered serous crime by removing signs of "disorder" through widespread police presence, frequent stop-and-frisks, and heavy emphasis on low level offenses and misdemeanour arrests. However, empirical evidence has consistently undermined this assertion (Harcourt, 2001). On the contrary, research has confirmed the tremendous violence Broken Windows policing has disproportionally caused in communities of colour over generations (Camp & Heatherton, 2016; Rios, 2011). Despite this, Broken Windows policing has remained the dominant framework in NYC for over twenty-five years and four mayorships.

Our initial maps of police activity citywide, by precinct, over time and by race, helped us as a PAR collective to enter a political conversation about public safety. Politicians, journalists, the NYPD commissioner, and other powerbrokers were using data curated by the police and from the police perspective to make strong public claims about the need for proactive policing and frequent stop-and-frisks to keep communities of colour safe. These policies were justified without those most directly impacted by them having meaningful representation in the debate. For the Morris Justice Project, rather than relying on maps produced by legitimised authorities – e.g. the NYPD, the mayor, the New York Times – we began a process of exploring for ourselves. Our participatory mapping of public data gave us the control to determine what and how we analysed, guided by the collective wisdom and interests of the group.

In an iterative and exploratory fashion, the Morris Justice collective used maps to reflect deeply on what it meant to live in a heavily policed neighbourhood of colour, and this guided us towards new analyses based on intimate local knowledge. For example, Morris Justice researchers expressed an embodied awareness of the cyclical patterns of police surveillance. The group told stories of changes in patrol by season, day of the week, time of day, Yankees home games, among other factors. This led us to create a timelapse map of every police stop in 2011 across NYC, spanning just over three minutes. Starting on January 1, 2011 and ending December 31, 2011, red dots quickly flashed in its geocoded location for every second, minute, hour and day over the entire year. Examining the spatial data in this way revealed a pulsing pattern across the city, with activity increasing towards the evenings. Additionally, for every flashed police stop that was innocent (no recorded arrest or summons), a permanent blue dot remained and accumulated, revealing what we called the "scars of our city," a pattern of relentless and lasting harm unevenly distributed across the geography. This work began as an internal analysis, but like many of our maps, it became a tool for advocacy as well. (For a video of the timelapse map, see Lizama, 2013).

The research camp provided space to reflect on and define the problem. It was also space to learn intimately with and from each other. Rather than a research project that explored NYC widely, we wanted to remain grounded in and accountable to a meaningful area. The strength and quality of our research as well as its liberatory possibilities rested on the situated knowledge that came from living in *this* South Bronx neighbourhood. To members of the Morris Justice Project who lived in the South Bronx, home was not defined by their precinct or census boundaries, nor was it defined by the trendy neighbourhood names designated by the gentrifying interest of developers. Defining the boundaries of home was a central task for our research project because it forced us to answer the critical question, whom are we representing and therefore accountable to?

We used South Bronx maps from Google of various sizes and configurations to facilitate a conversation about community boundaries. The meaning of home and therefore the boundaries

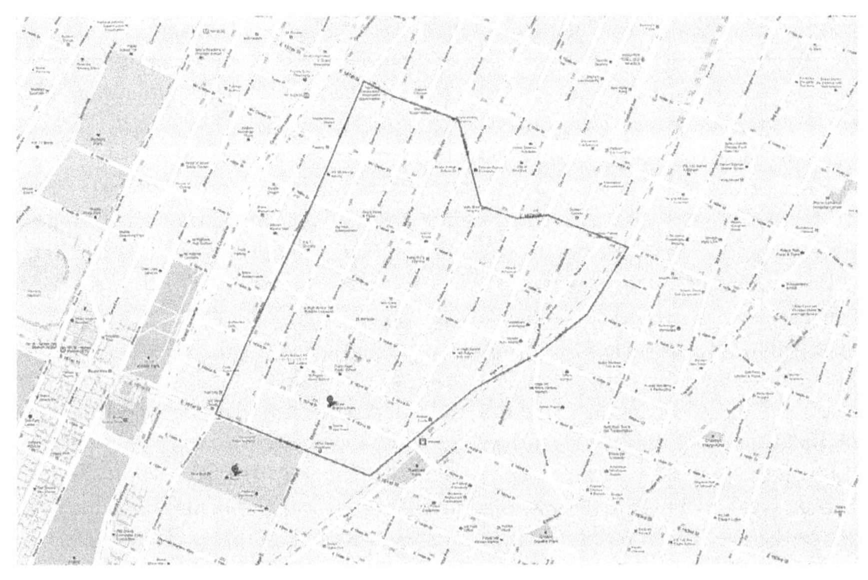

Figure 5.2 Top: timelapse map that grew from living spatial patterns of surveillance. Bottom: spatial reflection facilitated by a series of blank Google maps that helped us decide on the 42 square blocks the project called "home."

of *our* neighbourhood were different for each member who lived in the South Bronx, but over time and through rich discussion, we eventually agreed upon 42 square blocks that absorbed parts of the 44th and the 42nd precincts (see Figure 5.2). The right border followed the tracks of an aboveground commuter rail and created a physical barrier between community life. The left border purposely excluded Yankee Stadium and the Bronx Museum, as those did not feel like their neighbourhood, despite often defining it to outsiders such as tourists. The top and bottom borders involved a set of public housing complexes and larger (busier) streets that divided this neighbourhood from others. Community boundaries also excluded certain gang territories that felt dangerous to include for some of the young men in the group. For example, there was a church directly below our boundary that allowed us to meet, but the young men would not go there. Cutting down the middle of our 42 blocks ran Morris Avenue, where the library we regularly met at was located, hence how we became known as the Morris Justice Project.

Our goal was to analyse the stop-and-frisk data in the most meaningfully local way; to represent as best we could through public data the community impact of this NYPD policy. With our newly defined boundaries of home, we sought to explore yearly patterns of police stops *within* our specific neighbourhood blocks. The first attempt at this involved counting stops within census blocks and then writing by hand the totals on a large map, as well as marking the locations of stops that led to the confiscation of a gun. Through this neighbourhood-focused exploration, we learned that police recorded 3,920 stops in 2011 and only recovered eight guns. In comparison, buyback programs could collect ten times that number of guns in a single day. This type of spatial analysis was significant for the group. At once, it confirmed what they were intimately living, while revealing in very local terms the lie they were told, which was that police stops were a necessary and successful strategy for getting guns off the streets (see Figure 5.3). The numbers that produced this map would hold a prominent place in our spatial thinking; they were used for a t-shirt slogan and would even become the source of a winning lotto ticket. Yet for us, participatory mapping was an iterative process of exploring, reacting, and revising. Morris Justice researchers would eventually come to feel that while the map was illuminating, it fell short of representing home.

The group's critique of this map was born from their intimate spatial knowledge of walking through the neighbourhood. The problem with numbers aggregated by census block was that they did not reflect the situated experiences of living in a heavily policed space. While stops certainly happened inside hallways, rooftops, and courtyards, they predominantly occurred on street corners and sidewalks. From the perspective of someone sitting on their apartment stoop, census blocks arbitrarily cut the street down the middle so that police encounters on the sidewalk immediately in front, as well as the street behind, above, and below are aggregated into a single value. However, a stop that occurs directly across the street is aggregated into an entirely different census block. Our original map reduced, distorted, and sanitised police-resident interactions into aggregated counts floating in the middle of a square block, abstracted from the lived reality of urban street life and police presence. To stay consistent with our ethical commitment to see police engagement from the eyes of community members rather than through the state's eyes, we revised the map the next year to include counts by street and intersection, NOT by block. Our new map brought into focus and made abundantly clear that stop-and-frisks were much more prevalent on some streets than others (Figure 5.3).

Defining the Morris Justice neighbourhood allowed us to focus our research analysis on how our South Bronx members spatially defined *their* home. However, my home was in the East Village of Manhattan, near New York University, a similarly sized area in square miles and population (as I understood and lived it), but a majority-white and generally affluent neighbourhood. Most, but not everyone, in our collective lived within or near the boundaries of the Morris Justice neighbourhood. Some, like me, presented important distinctions to unpack. For example, I could sit outside in the nearby park with a bottle of wine without fear of police harassment, yet merely drinking a beer on one's own stoop or courtyard in the Morris Justice neighbourhood could elicit a stop-and-frisk. The juxtaposition of our defined South Bronx and East Village boundaries became a significant comparative inquiry for us. We discovered that the East Village had 2,747 fewer stops in a year; 2,938 fewer frisks; 440 fewer searches; and 2,340 fewer stops resulting in physical force. Yet the East Village also had 72 *more* arrests with only 1 less

Police Stops & Gun Confiscation in 2011: Block By Block
3,920 total stops (on average 11 per day) for 8 guns

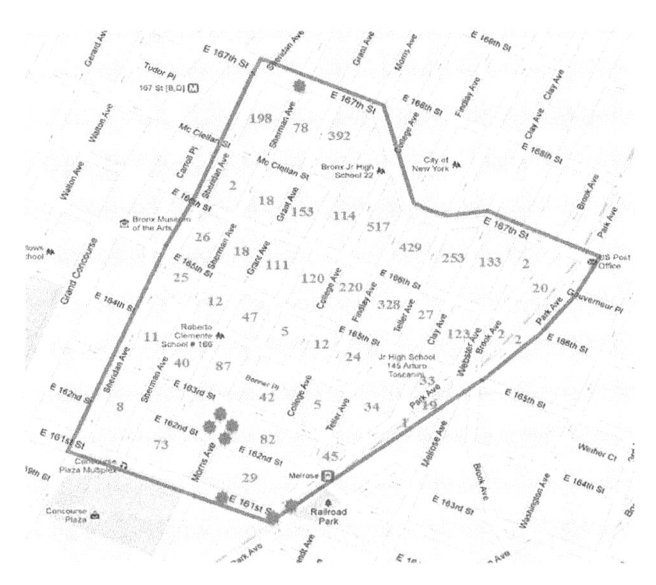

In 2012, there were 4,037 police stops in this neighborhood (on average, that's 11 stops per day) FOR ONLY 7 GUNS

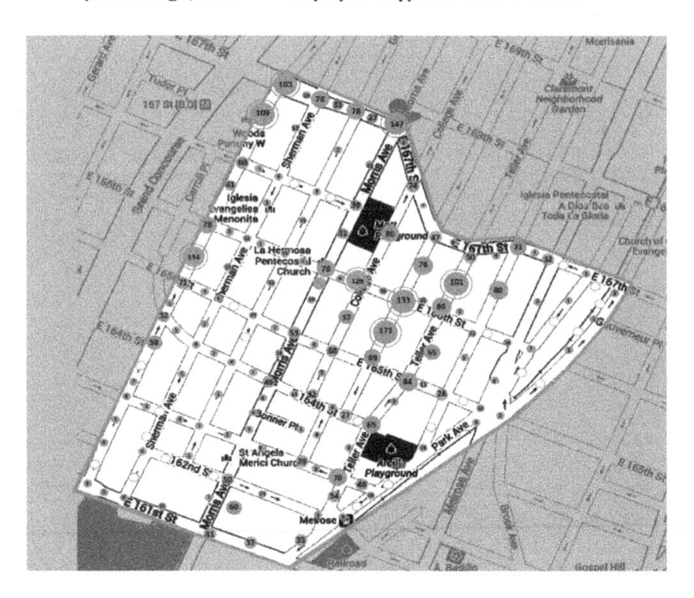

Figure 5.3 Top: total police stops in the neighbourhood by census block with red dots displaying gun recovery. Bottom: the transition to mapping the lived experience of stops, displaying totals by streets and intersections.

gun, 2 less knives, and 55 *more* contraband recovered. Our analyses revealed a "tale of two neighbourhoods," clear evidence of the spatial bias of carceral surveillance in communities of colour (see Figure 5.4).

A Tale of Two Neighborhoods

Using 2011 NYPD Stop, Question, and Frisk data and 2010 US CENSUS DATA

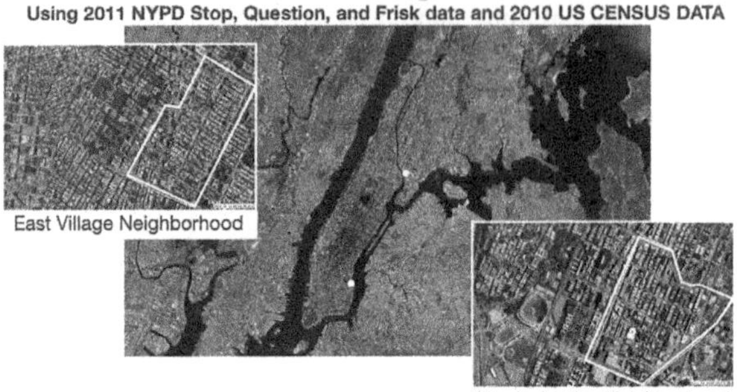

East Village Neighborhood

Morris Justice Neighborhood

Residents in the Morris Avenue neighborhood are policed differently than residents in a section the same square miles and similar population size in the East Village. In fact, *The New York Times* reported that our precinct (the 44th) and the neighboring 46th "use physical force far more often than the police do anywhere else in the city" (August 15th 2012). **A comparison of how our 40 blocks just east of Yankee stadium are policed versus a similar area in the East Village just east of NYU shows how unequal things are depending on where you live and who you are.**

East Village Neighborhood (.243 sq mi. Pop. 22,889) Part of 9th precinct 68% White		**Morris Justice Project Neighborhood** (.239 sq mi. Pop. 19,173) Part of 44th precinct 96% Black/Latino(a)
2135	Total police stops	4882
45.5% (971)	Stops involving frisks	80.1% (3909)
12.7% (271)	Stops involving searches	14.6% (711)
81.9% (1748)	Stops NOT leading to arrest or summons (Innocent)	92.5% (4517)
.23% (7)	Stops leading to gun recovery	.12% (8)
1.8% (38)	Stops leading to knives, cutting instruments, and other weapons recovered	.82% (40)
6.1% (130)	Stops leading to contraband recovery	1.5% (75)
24.3% (519)	Stops involving physical force	58.6% (2859)
68.8% (357)	Percent of physical force stops not leading to arrest or summons (innocent)	90.8% (2597)
15.8% (338)	Stops leading to arrests	5.4% (266)
2.3% (50)	Stops leading to summons	2.4% (118)
16.5% (352)	Stops on people 20 years old or younger	42.2% (2066)

Figure 5.4 Our spatial analysis of public data from police stops comparing the Morris Justice and East Village neighbourhoods

In this section, I describe how the Morris Justice Project used participatory mapping to learn about and reframe understanding of the neighbourhood. Our iterative spatial analyses of public police data, mapped at multiple scales and with different comparisons, were threaded throughout our research and continually optimised to illuminate local conditions. It empowered members of the Morris Justice Project to co-construct intimate knowledge *about* and hold ourselves accountable *to* the 42-blocks in the South Bronx they called home. In the next section, I move to our community-based survey and the subsequent community mapping processes we implemented to partner *with* the neighbourhood to understand home.

Mapping *with* Home

Determining the boundaries of what we came to understand as *our* Morris Justice neighbourhood helped us to spatially focus our research camp discussions with questions such as, What is it like to live in this community? We made large maps of the 42 blocks (see Figure 5.5), then wrote directly onto the posters as a method to understand how people used the neighbourhood.

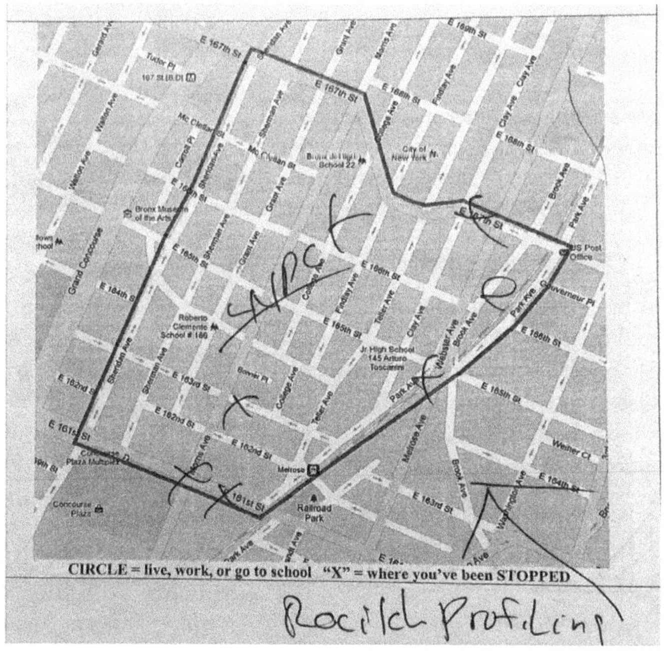

Figure 5.5 Top: Morris Justice researchers writing directly on a large map of the neighbourhood to ground local expertise and experience. Bottom: an example of spatial data we collected by incorporating a mapping exercise in our community survey

Where did they and their friends live? Where was school or work? Where did they shop or get a haircut? Where did they have fun? Where were they stopped by police? Where to expect heavy police presence? It was here we started theorising how police patrolled streets differently and noticed how some members had multiple police contact scattered across the blocks. These spatial strategies helped to unpack our assumptions and created space to name the challenging, amazing, and ambiguous parts of living in the neighbourhood (Brewster, Encandela, Stoudt, & Fine, 2014).

Our maps of public NYPD data reduced space and the people living in that space down to a dot or an aggregated number. Though illuminating, the Morris Justice collective wanted to connect more closely with neighbours to find out what *their* experiences were and how *they* understood police. It was in these moments of defining the community boundaries, subsequently reflecting on conditions of home, and recognising a deeper desire to talk with residents, that early conceptual domains and questions emerged. These would become a set of interviews, focus groups, and particularly a community survey. As our community survey took shape, we felt it necessary to incorporate a spatial analysis by asking residents in the Morris Justice neighbourhood to map their own experiences with police. This was done by marking the locations of stops and the corner nearest to where they live, work, or go to school (see Figure 5.5). By mid-fall, as the research camp concluded, the Morris Justice collective had spent many hours across many drafts integrating our collective assumptions, theories, and questions into the survey. Now finalised, it was time to learn directly from neighbours.

Distributing the survey was an opportunity to hold meaningful conversations about policing with members of the community. This meant developing a viable community-grounded sampling strategy where we could take/hold space for these social interactions (Stoudt, 2016). Although a small area, these 42 blocks represented enclaves of varied and vibrant lives that we felt accountable towards. It was important to hit every corner, so all residents had an opportunity to take the survey. Once again, our maps organised our actions, helping us remain rooted and systematic. The primary sampling method involved cutting the neighbourhood's 42 blocks into 18 smaller sections, each section including between one to three blocks (see Figure 5.6). Using clipboards, a stack of surveys, sharp pens, stickers advertising the project, and roundtrip subway cards as incentive, the research team scattered around the designated blocks within the section assigned that day. The research team methodically spent time on every street in the map, collecting 1,030 surveys in both English and Spanish over six months (see Figure 5.6). These surveys counted 1,424 police stops in the previous year.

It was clear from our earliest discussions that young people involved with the Morris Justice Project had frequent contact with police over the previous year. However, we were unable to investigate the extent of these trends using public NYPD data, because it was organised by incidents without reference to individuals. One aim of our community survey was explicitly to address this gap, and it revealed that 69% of the respondents reported they were stopped at least once in the last year. Of those who were stopped, 82% were stopped more than once, and 52% were stopped four or more times. Some of the young men in our project also felt that they and their friends were targeted for living in buildings with bad reputations. Our group theorised where spatial clusters of multiple stops were likely to occur, and used the maps drawn on the survey by respondents to explore those trends further.

We developed a participatory GIS strategy to examine clusters of multiple stops. We produced five maps that displayed where people in the neighbourhood lived who had been stopped 1+, 2+, 5+, 10+ and 20+ times in the last year. The group worked alone

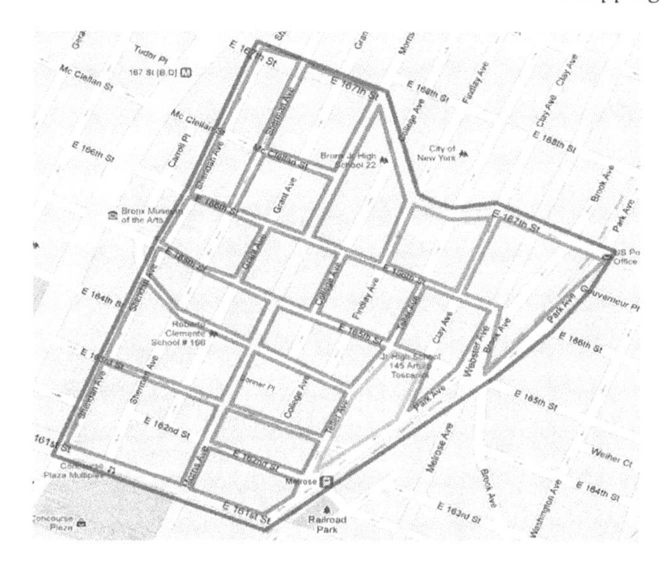

Morris Justice Project Community Survey

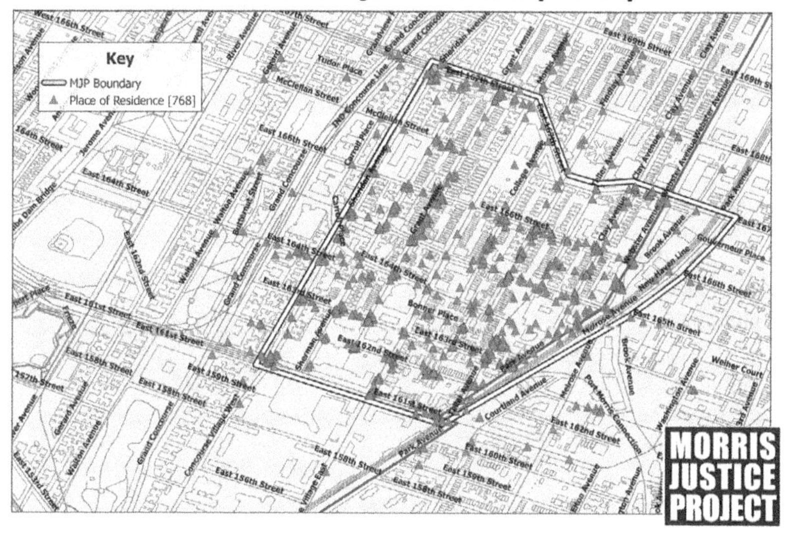

Figure 5.6 Top: the 42 blocks of the Morris Justice neighbourhood were cut into 18 smaller blocks to systemically sample. Bottom: A GIS map constructed to represent where the respondents of our community survey lived, worked, or went to school

and in small groups to scan the maps closely and circle clusters they felt existed (see Figure 5.7). A commonly recognised pattern among group members involved three clusters, two in the upper part of the neighbourhood, and one in the lower right corner, particularly in the maps displaying people who were stopped 10+ and 20+ times (See Figure 5.7). The lower right corner involved buildings of several Morris Justice members and confirmed their hunch. The other two clusters led us to consider the ways that police were surveilling several homeless shelters and a local park known for drug activity.

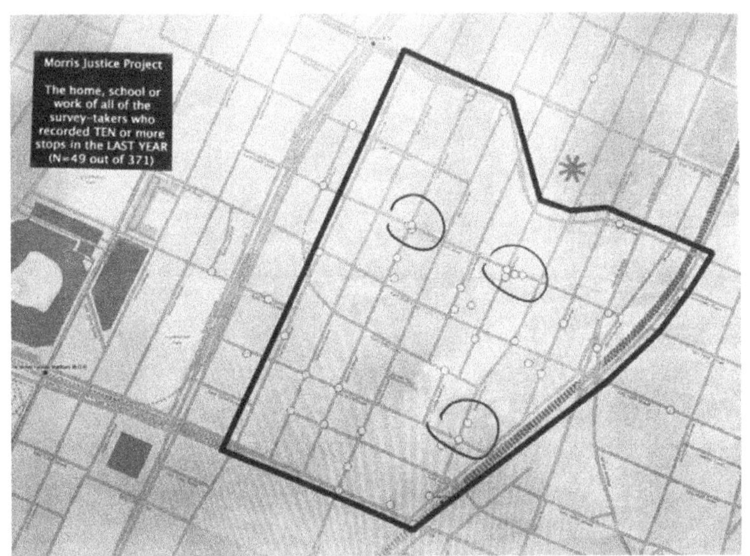

Figure 5.7 Top: Morris Justice researchers doing a participatory mapping exercise to explore residential patterns from the community survey of multiple stops in the neighbourhood. Bottom: an example of spatial clusters that emerged and represented heavy police surveillance

Each summer/fall, the Morris Justice Project (re)claimed our "right to the city" (Mitchell, 2003) through a process called "sidewalk science" (Stoudt et al., 2019; Torre et al., 2018). This was a series of interactive installations and stations held impromptu on the streets, both to share results and to collect new data. Interactive community maps were central components of sidewalk science sessions, which we used as strategies to spatially organise dialogue among visitors and unpack tensions revealed in our research. One technique involved hanging a large neighbourhood map on corner fences or walls and encouraging people passing by to use yellow sticky notes to locate and describe places where they felt a sense of community (see Figure 5.8). Their responses represented the ordinary ways that people make and remake space and place every day: in "church," "my sister's place," and "with my family." They

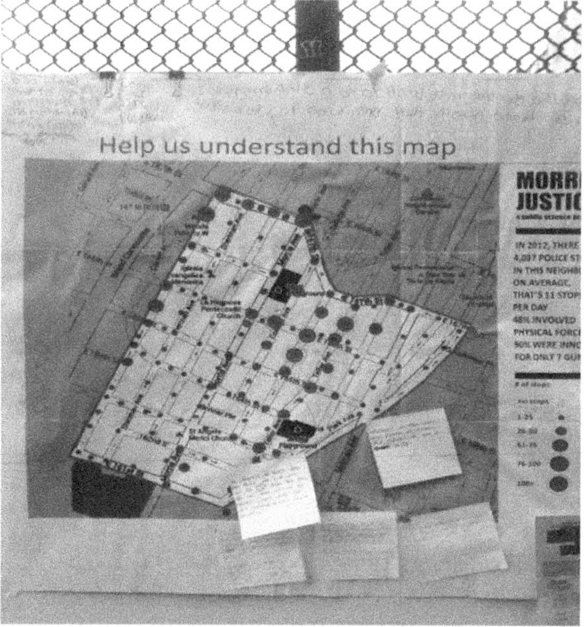

Figure 5.8 Top: a sidewalk science event using a participatory mapping activity to explore a sense of community and police disruption. Bottom: A sidewalk science event using a participatory mapping activity to understand the spatial trends of NYPD police stops

explained relationships: "I love the people in it," with "my new friend," "people that have been there a long time," or among familiar strangers at establishments like "Kennedy fried chicken – they are very respectful + nice." Their description of community also represented the long history of divestment in their neighbourhood and South Bronx generally: "Good people but no activity for kids. Need a playground." The absence of spaces and activities for young people was a common theme throughout our research.

The map also instructed visitors to use red sticky notes to locate and describe places where police disrupted their (sense of) community (see Figure 5.8). One person placed a yellow sticky note over a building and explained "I have a lot of friends there," and below it placed a red sticky note that said, "but there's also a lot of police there. Too many." Police surveillance can quickly become police contact: "[I was] stopped for no reason. Said my friend looked suspicious." Living in a heavily policed area means that making community – visiting friends and family, walking outside, or sitting on a stoop – is also inseparably coupled with negotiating the threat of police contact and possible escalation. The consequences of that threat are serious: "[I] was stopped for no reason. I was taken in b/c my friend had a small joint. Went through the whole system. I almost lost my job and now I don't bring my kids out anymore." This map created a public gallery, a growing archive of community and policing that others who walked by could reflect on and use to engage in critical conversations. This was a moment to discuss the role of police, one's sense of the neighbourhood, and what residents desire in the place they call home. One person even wrote of the sidewalk science mapping, "This right here. This makes me feel connected + like there is hope."

Another community mapping exercise for sidewalk science involved enlarging our spatial analysis of recorded police stops throughout neighbourhood streets and intersections. A poster provided a short introduction with the instructions "Help us understand this map" written on the top. Those passing by were encouraged to write directly on the map or post a sticky note (see Figure 5.8). People drew, wrote, starred, put arrows, and gave explanations. One young man of colour took a picture and sent it to his friends, left, and returned with more friends. He told us it was interesting for him to see all the stops made in the areas he spent time. He also said it validated his own experiences with police and offered his own interpretation of the spatial trends. Other did too. Some proposed theories about the stops: "Majority of the stops are Black and Brown people between the age of 16–25" or "they are being targeted." Others recognised themselves or their friends represented on the map, and wrote what happened: "stopped and arrested for pocket knives" or "harassment." A few responded with questions like, "How many young people are being stopped and frisked just for the way they look, dressed and who they hang out with?"

Community mapping activities like this brought neighbours into public conversation, linking in solidarity a diversity of perspectives, people, ages, and experiences. There was power in presenting meaningfully local data on a large map. There was power in offering residents the tactile control of writing directly on the map. These were brief moments of community interaction that offered a space to express their experiences and expertise. One example stood out and involved residents who questioned the validity of the mapped NYPD numbers: "How many stops are actually recorded. This may look like a lot of stops to you, but I think this is low?" Another wrote, "We believe stops are not being recorded." A third drew an arrow pointing to two parts of the neighbourhood: "There's no way there's more stops here [147 at one street corner] than here [129 at another street corner]." A few years later, because of winning three lawsuits against the NYPD, the general public would learn what many in this neighbourhood already understood: the police were in fact, substantially under-recording stops.

In this section, I describe how the Morris Justice Project spatially analysed experiences with police in conversation with the neighbourhood through community surveys and sidewalk science activities to understand home with nuance and complexity. This process was a way to directly speak with community residents and centre their situated expertise. It served as a snapshot of police harassment but also a mirror into their own daily lives, intimately tied to the carceral state. Mapping with the community revealed how vital geographic understanding was in relation to police surveillance. The maps created debate, a chance to (re)see patterns and then theorise them. People expressed feeling that their individual experiences were validated and contextualised, revealing a sense of mutual implication and connection. Participatory mapping continually held the project accountable to the local, offering pathways for intentional enactments of spatial commitments. In the next section, I further explore the concept of critical accountability structures.

Critical Accountability Structures, Spatial Accountability and Participatory Mapping

For whom is research accountable? The Morris Justice Project held itself accountable to each other, as a diverse collective of community-based and academic researchers, and to the residents who lived, worked, and/or went to school in the designated 42 blocks that many of our members understood as home. However, for whom we are accountable is different than *how* we are held accountable. The former involves intention, the latter involves intentional practice, deliberately enacting guardrails, guideposts, and constraints through designs, methods, and practices to invite and co-construct accountability. Accountability structures – how research is held accountable to interests, values, commitments, and perspectives – are part of all research, though seldom made explicit and frequently taken for granted.

Consider some of the accountability structures of generic academic research. A project might start out accountable to funders (or what is fundable), and to an established body of knowledge, as well as scientific norms. In practice, the researcher might review relevant academic scholarship to theorise the inquiry, develop research questions, and craft an appropriate design (e.g. cross-sectional) with standardised methods (e.g. survey). The systematic practice of gathering and making meaning of evidence is also a process of accountability, constrained by the sampling procedures, the targeted sample, the information individuals provide, and the techniques of analysis. Indeed, a great value of research is the commitment to systematic data collection and examination that incorporates a diversity of thought and experience beyond one's own. Communicating findings in a journal through the scrutiny of peer-review is an additional process of scholarly accountability, often considered the gold standard of rigor.

Conducting research from start to finish involves a complicated network of accountability practices – relationships, commitments, rules, expectations, and decisions – that offer constrained and disciplined pathways to frame, focus, open, recognise, understand, and act. Accountability structures offer potentials, though not necessarily less distorted potentials. Even when researchers hold the best intentions and produce outcomes that are of the highest quality (by academic standards), it is easy to see how well-worn accountability practices create a gravitational pull towards the privileged interests of the academic community, only accentuated further when accounting for the boundaries of professional organisations, the standards of tenure/promotion committees, and the politics of departments/universities. Of course, the participatory action research literature has long recognised and critiqued the narrow elitism of academic knowledge production and offers a diversity of counter-practices that decentre the hegemony of academic accountability (Glass & Stoudt, 2019).

PAR attempts to revise what counts as knowledge, expand who counts as expert, decolonise methodology, and pursue liberatory action. It seeks practices that centre the experiences, interests, and wisdom of those most impacted by structural oppression. It values diverse viewpoints, intersectional complexity, intimate relationships, and political solidarity. It is steeped in the emancipatory and ethical framings of critical race, queer, feminist, and postcolonial theories. It rejects methodological extraction and exploitation, instead positioning those historically marginalised from legitimised knowledge production with the means to pursue meaningful and actionable research. These are all ongoing and intentional enactments that participatory action researchers must make real through methods and practices that structure a set of relations to hold them accountable for these commitments. PAR revises, expands, and often queers mainstream research accountability practices. Indeed, from the perspective of a research trajectory that includes theorising inquiry, designing research, gathering evidence, making meaning, and communicating findings, PAR invests in a more robust, more purposeful, more expansive set of *critical accountability structures* across each of those phases.

The Morris Justice Project serves to illustrate this point. We were a team of co-researchers, a fact true of many research projects, *and also* our collective was mostly people of colour directly impacted by aggressive policing, with little to no previous background in research. We drew from the policing literature across a host of social science disciplines *and also* from relevant research reports produced by grassroots and community-based organisations. We co-constructed our research questions, produced a survey, and developed interview questions all grounded in the interests and experiences of South Bronx residents, *and also* we created a community advisory group for additional feedback. We systemically sampled using our methods to elicit information within the meaningful boundaries of our neighbourhood, *and also* we organised this experience as a series of community conversations (rather than a process of sanitised extraction). Together, we quantitatively and qualitatively analysed our data, *and also* we took our preliminary findings into the neighbourhood to elicit additional assistance with our interpretations, a practice we called "community analysis." We communicated our findings through co-written reports and peer-reviewed articles, *and also* distributed our findings back to the neighbourhood through sidewalk science and the citywide police reform movement (see Figure 5.9). Accountability is a characteristic of intentionally organised group dynamics. In each phase of our research, we deliberately sought practices that compelled us, as much as possible, to implement our commitments.

This chapter documented the Morris Justice Project's pursuit of spatial understanding and accountability to push against carceral violence. It demonstrated innovative mapping

Figure 5.9 Left: South Bronx residents of the Morris Justice Project presenting our participatory mapping analyses to an academic audience. Right: Member of the Morris Justice Project using the Illuminator to locally present our research findings on a busy street corner in the South Bronx

strategies that were informed by a participatory approach to statistics (stats-n-action) grounded in the liberatory possibilities of exploring localised numbers to generate reflexive dialogue to learn, theorise, and act spatially. I explored how participatory mapping – as design, method, and action – was intentionally woven into the fabric of our research-action trajectory to stay firmly rooted in 42 blocks of the South Bronx. This was an ethical commitment to *home* that we understood as imperative to the validity of our work. I specifically outlined, within the context of the Morris Justice Project, how the ongoing use of participatory mapping took on two emergent pathways. The first section illustrates the *internal practices and processes* of our collective as we iteratively attempted to understand and reframe the prevailing messages produced from the publicly available NYPD stop-and-frisk data. It shows how collective members drew upon their locally situated knowledge to reflect critically and make spatial meaning. The second section demonstrates the *outward practices and processes* of our collective as we oscillate between the internal survey work of the group and the deliberate consultation with the neighbourhood. It shows how the spatial analyses occurring within the Morris Justice Project are continually extended out into the neighbourhood, creating an ongoing spatially situated conversation between our collective and its residents.

Critical accountability structures serve as a lighthouse, guidepost, and guardrail for participatory action researchers. The lighthouse keeps commitments in focus, the guidepost offers possible paths, and the guardrail constrains when principles wane. Participatory mapping was an essential method for the Morris Justice Project because it continually compelled us to stay rooted in the knowledge of home, while creating a context to collectively challenge the acute violence of the carceral state.

***

On a brisk September evening on the corner of a busy intersection, the Morris Justice Project partnered with an activist art group called the "Illuminator" that used a large projector fastened to the roof of their van to beam our research findings onto the side of a 20-story public housing building (See Figure 5.9; for a video of the event, see The Illuminator, 2012). We presented our work in the form of an open letter to the local NYPD who patrolled the South Bronx. Over the loudspeakers members of the collective who lived in the neighbourhood took turns reading our co-written script to the gathering crowd. Each research finding was followed by a refrain referencing home. The refrain was a poem, the result of deep spatial analysis through participatory mapping, asserting their right to the city and asking the police for dignity and respect in the place they call home:

Dear NYPD,

WE ARE the Morris Justice Project
This is our home.
WE. LIVE. HERE.
We BELONG here
Please don't treat us like we're strangers
This is our home
We go to school here
Please don't assume we're criminals.
This is our home
We raise our kids here
Please don't physically abuse them
This is our home
We play here
Please don't tell us to move when we're not doing anything wrong
This is our home
We shop here
Please don't stop us without a good reason
This is our home
We pray here
Please treat us with dignity and respect
This is our home
We care about our neighbourhood
We want to feel safe
And we want you here
Dear NYPD
We ARE the Morris Justice Project
We DESERVE fair and just policing

Acknowledgements

This work was made possible through the generous support of The Public Welfare Foundation, Open Society Foundation, and The Public Science Project. A special thank you to Rachel Pain for her amazing editorial feedback. And a very special thanks to María Elena Torre and all my co-researchers at Morris Justice Project

References

Bradshaw, E. A. (2013). This is what a police state looks like: Sousveillance, direct action and the anti-corporate globalization movement. *Critical Criminology*, 21(4): 447–461. https://doi.org/10.1007/s10612-013-9205-4.

Brewster, K., Encandela, J., Stoudt, B., & Fine, M. (2014). Experience and domain mapping as dynamic tools for focus groups. *Social Work Research*, 38(3): 184–189. https://doi.org/10.1093/swr/svu008.

Brickell, K. (2012). Mapping and doing critical geographies of home. *Progress in Human Geography*, 36(2): 225–244. https://doi.org/10.1177/0309132511418708.

Bunge, W. (2011). *Fitzgerald: Geography of a Revolution, Vol. 8*. Athens, GA: University of Georgia Press.

Camp, J. T., & Heatherton, C. (Eds.) (2016). *Policing the Planet: Why the Policing Crisis Led to Black Lives Matter*. London: Verso Books.

Chambers, R. (2007). Who counts?: The quiet revolution of participation and numbers. Institute of Development Studies. https://www.participatorymethods.org/.

Coates, T. (2013, 25 July). The dubious math behind stop and frisk. *Atlantic Monthly*. https://www.theatlantic.com/national/archive/2013/07/the-dubious-math-behind-stop-and-frisk/278065/.

Cochrane, L., & Corbett, J. (2020). Participatory mapping. In J. Servaes (Ed.), *Handbook of Communication for Development and Social Change*. 705–713. Singapore: Springer.

Crampton, J. W. (2011). *Mapping: A Critical Introduction to Cartography and GIS, Vol. 11*. Hoboken: John Wiley & Sons.

Dalton, C. M., & Stallmann, T. (2018). Counter-mapping data science. *The Canadian Geographer/Le Géographe Canadien*, 62(1): 93–101. https://doi.org/10.1111/cag.12398.

Desrosières, A. (1998). *The Politics of Large Numbers: A History of Statistical Reasoning*. Cambridge, MA: Harvard University Press.

Du Bois, W. E. B. (2007). *The Philadelphia Negro*. New York: Cosimo, Inc.

Eterno, J. A., & Silverman, E. B. (2012). *The Crime Numbers Game: Management by Manipulation*. Boca Raton: CRC Press.

Geller, A. (2015). The process is still the punishment: Low-level arrests in the broken windows Era. *Cardozo Law Review*, 37: 1025–1026. http://cardozolawreview.com/.

Giuliani, R. W., & Bratton, W. J. (1994). *Police Strategy No. 5: Reclaiming the Public Spaces of New York*. U.S. Department of Justice. https://www.ojp.gov/.

Glass, R. D., & Stoudt, B. G. (2019). Collaborative research for justice and multi-issue movement building: Challenging discriminatory policing, school closures, and youth unemployment. *Education Policy Analysis Archives*, 27(52). https://epaa.asu.edu/

Harcourt, B. E. (2001). *Illusion of Order: The False Promise of Broken Windows Policing*. Cambridge, MA: Harvard University Press.

Hunt, D., & Stevenson, S. A. (2016). Decolonizing geographies of power: Indigenous digital counter-mapping practices on Turtle Island. *Settler Colonial Studies*, 7(3): 372–392. https://doi.org/10.1080/2201473X.2016.1186311.

Irvine, J., Miles, I., & Evans, J. (1979). *Demystifying Social Statistics*. London: Pluto Press.

Kidd, D. (2019). Extra-activism: Counter-mapping and data justice. *Information, Communication & Society*, 22(7): 954–970. https://doi.org/10.1080/1369118X.2019.1581243.

Lizama, S. (2013, 20 March). *This Is Our Home*. https://youtu.be/qLWWa2De2b4.

Maharawal, M. M., & McElroy, E. (2018). The Anti-eviction mapping project: Counter mapping and oral history toward Bay Area housing justice. *Annals of the American Association of Geographers*, 108(2): 380–389. https://doi.org/10.1080/24694452.2017.1365583.

McCall, M. K. (2006). Precision for whom? Mapping ambiguity and certainty in (participatory) GIS. *Participatory Learning and Action*, 54(1): 114–119. https://pubs.iied.org/.

Mitchell, D. (2003). *The Right to the City: Social Justice and the Fight for Public Space*. New York: Guilford Press.

Muhammad, K. G. (2019). *The Condemnation of Blackness: Race, Crime, and the Making of Modern Urban America, with a New Preface*. Cambridge, MA: Harvard University Press.

Rios, V. M. (2011). *Punished: Policing the Lives of Black and Latino Boys*. New York:NYU Press.

Ritchie, A. J. (2017). *Invisible No More: Police Violence Against Black Women and Women of Color*. Boston: Beacon Press.

Schneiderman, E. T., NYS Attorney General, & Civil Rights Bureau (2013). *A Report on Arrests Arising from the New York City Police Department's Stop-and-Frisk Practices*. New York State Office of the Attorney General. https://ag.ny.gov/.

Smith, L. T. (1999). *Decolonizing Methodologies: Research and Indigenous Peoples*. London: Zed Books.

Stoudt, B. (2014). Critical statistics. In T. Teo (Ed.), *Encyclopedia of Critical Psychology*. 1850–1858. New York: Springer-Verlag.

Stoudt, B. G. (2016). Conversations on the margins: Using data entry to explore the qualitative potential of survey marginalia. *Qualitative Psychology*, 3(2): 186–208. https://doi.org/10.1037/qup0000060.

Stoudt, B. G., Fine, M., & Fox, M. (2011). Growing up policed in the age of aggressive policing policies. *New York Law School Law Review*, 56(4): 1331–1372. https://www.nylslawreview.com/.

Stoudt, B. G., Torre, M. E., Bartley, P., Bissell, E., Bracy, F., Caldwell, H., *et al.* (2019). Researching at the community-university borderlands: Using public science to study policing in the South Bronx. *Education Policy Analysis Archives*. https://epaa.asu.edu/.

Stoudt, B. G., Torre, M. E., Bartley, P., Bracy, F., Caldwell, H., Downs, A., Greene, C., Haldipur, J., Hassan, P., Manoff, E., Sheppard, N., & Yates, J. (2015). Participatory action research and policy change. In C. Durose & L. Richardson (Eds.), *Designing Public Policy for Co-Production: Theory, Practice and Change*. 125–137. Bristol: Policy Press.

The Illuminator (2012, 22 September). *Stop and Frisk in the South Bronx*. https://youtu.be/mliuISC2hJk.

Torre, M. E., Fine, M., Stoudt, B., & Fox, M. (2012). Critical participatory action research as public science. In P. Camic & H. Cooper (Eds.), *The Handbook of Qualitative Research in Psychology: Expanding Perspectives in Methodology and Design*. 171–184. Washington, DC: American Psychological Association.

Torre, M. E., Stoudt, B. G., Manoff, E., & Fine, M. (2018). Critical participatory action research on state violence: Bearing wit(h)ness across fault lines of power, privilege, and dispossession. In N. Denzin & Y. S. Lincoln (Eds.), *The SAGE Handbook of Qualitative Research*. 492–515. Thousand Oaks, CA: Sage Publications.

Tukey, J. W. (1969). Analyzing data: Sanctification or detective work? *American Psychologist*, 24(2): 83–91.

Vitale, A. S. (2017). *The End of Policing*. Brooklyn: Verso Books.

Zuberi, T., & Bonilla-Silva, E. (Eds.). (2008). *White Logic, White Methods: Racism and Methodology*. Landham, MD: Rowman & Littlefield Publishers.

6 Using Participatory Action Research for Performing Stories and Imagining Inclusive Communities

Nina Woodrow

In this chapter, I trace the development of a participatory action research (PAR) project that proceeded through the material practices of storytelling mediation. The Brave New Welcome (BNW) Project engaged with a cohort of over a hundred young people, including many from refugee backgrounds, in an urban setting in Australia. The project collective also included local arts workers, activists, peace building, and settlement support professionals. For this researcher/practitioner team, the underpinning intention was to launch a study that eschewed the trope of the refugee trauma story, centred on individuals with who needed "help" to integrate into a new social setting. The intention instead was to shift the focus away from the refugee as the "subject" of research. We wanted to learn something, in partnership with the young people involved in the study, about the role of the host community in staging a better welcome, in orchestrating better settlement experiences for young people from refugee backgrounds. In other words, we were looking for new insights into how to be better advocates and better hosts. The starting point was what we saw as a moral, political, and cultural deficit among the broader population of Australians to adequately respond to refugees' circumstances.

Using PAR for Performing Stories and Imagining Inclusive Communities

The purpose of this chapter is to show how this starting point led us to engage productively with different ways of capturing experiences using aesthetic tools, along with different ways of sharing both creative outcomes and knowledge with various audiences and stakeholders.

We found that while PAR formed the backbone of the study, drawing from methods of performative inquiry and public ethnography expanded our tool set. We grasped early that the scope of this project meant venturing into very complex ethical territory. Extending an aesthetic and playful antenna into the questions and problems that framed the study lent us extra space to move and learn. This space made it possible to frame "fieldwork" as a co-creative and performative practice. The exploration of collaborative methods described in this chapter highlights the creative complexities that characterise a storytelling strategy, an approach often employed by artists, activists, cultural producers, and scholars committed to using socially engaged methods. In this chapter, I focus on what we, as hosts, as professionals, and as cultural practitioners, learned about staging a better welcome.

The design of the BNW project was evolving and layered, progressing through action research cycles activated by socially engaged arts practices. PAR literature documents a vibrant tradition of exploring the intersections between PAR and arts-based methods (Hume-Cook et al., 2007; Raynor, 2019; Tolia-Kelly, 2007). We discovered, as have many others before us, that these cycles of participatory art, reflection, and planning permitted us to expand the

DOI: 10.4324/9780429400346-6

opportunities for the co-production of knowledge. This co-creation became a significant part of what we collectively learned and achieved. Like Gallagher, Wessels, and Ntelioglou (2013), we as researchers, artists, and activists invested in this project found that a combination of live and digital methods, woven into an ethnographic engagement with diverse groups of young people, can open more opportunities for foregrounding questions of power and voice.

The Brave New Welcome Project

In 2014, a group of twenty-five young people from diverse backgrounds came together to participate in a program of storytelling activities. These activities were initiated by a callout to young people in the local community interested in exploring better ways to welcome young people from refugee backgrounds, newly arrived in Australia. This group coordinated the project stages and cycles, with the support of a temporary but purposeful alliance of cultural practitioners – an agile team of arts-workers, researchers, settlement support workers, and peace-building professionals. The BNW project, in the end, was simultaneously a program of applied research and a campaign of performed activism. In addition to the academic outputs, the project generated several of the creative outcomes. These artefacts, produced through a collaborative process (a short film and collaborative artwork) – along with the events orchestrated to stage and screen these outcomes – are a parallel, collective form of reporting on the research. These forms of public engagement aligned with the activist intentions of the whole project team in this community-based setting.

The BNW project developed through a series of meetings, workshops, and events over two years. The project emerged from an initial intention to engage with young people from refugee backgrounds living in the local Brisbane community, using storytelling and participatory media approaches. It started as a pilot digital storytelling project, with a small group of newcomers who had recently settled in Australia and were living on the outskirts of Brisbane. This group of storytellers shared tales of their dreams and ambitions through participatory arts and media activities. As we concluded, we found that a striking theme of social exclusion had emerged. Through the pilot phase of sharing lived experiences and stories of social isolation, via co-creating and sharing the collection of digital stories, a kernel was revealed. The whole project team took this kernel through four more cycles of project planning, action, and reflection. These phases of activity led to more ambitious, complex outcomes and to a new hybridised "co-performative" practice, which drew on the skills of a multidisciplinary facilitating team and the creative renderings of the storytellers.

By the second action phase, the planning moved on from its focus on young refugee-background people to our intention to bring a diverse group of young people together, including refugee-background and Australian-born alike. The aim here was to provide a supported environment and a context where young people who would not normally have contact could meet each other, build relationships, and explore ideas and responses to the theme of "welcome." Project planning was focused on supporting this process through arts-based practices, such as community theatre and participatory media. At this point, the project storytellers became known as "Brave New Welcomers."

In a week-long creative development intensive in January 2014, the Brave New Welcomers engaged in a multi-arts program (see Figure 6.1). This culminated in a day of facilitated dialogue based on the World Cafe method (Brown, Homer, & Isaacs, 2007). The group included a cluster of very recently arrived young people from refugee backgrounds, several young people from refugee backgrounds who had come to Australia as

Figure 6.1 BNW Creative Development Week at The Edge, January 2014

children or infants, some Australian born young people who were the children of migrant parents, as well as several second-generation or more Australian young people. This was a newly formed group of twenty-five young people with not much in common at the outset, apart from their age. For the facilitating team, the aim was to support these young people to share stories in creative ways, to build relationships.

The arts program included a range of artistic and dramatic group activities. With the direction of a skilled "playback theatre" practitioner (Fox, 2007), the group was supported through a series of theatrical exercises to turn each other's stories into "instant theatre." These story sharing activities, inspired by playback theatre, were also supplemented by visual arts and theatre games. This range of activities aimed to support this diverse group of young people to build trust through discovering a shareable aesthetic language.

As an outcome of this week, the Brave New Welcomers and the team of facilitators collectively decided to put two new plans into action. The first was to invite a wider group of local young people to participate in the discussion about friendship, and what a good welcome looks like, by hosting a public forum. The second was to embark on the co-creation of a film to document these embodied explorations of what makes a good community and how a good welcome could be practiced. The script for the film was developed collaboratively and was written in two stages by two groups of young people. The original Brave New Welcomers who participated in the initial creative development week wrote the first stage. The second stage was written by a bigger group of about a hundred young people who attended the BNW Youth Forum in May 2014 (see Figure 6.2). The original group produced the first part of the script and worked with a filmmaker to create an intentionally *unfinished* short film. The script was completed after the forum, with the input of the larger group, and the short film was finalised in time to be screened at the Brisbane celebration of the Refugee Film Festival in June 2014 held at the State Library (Woodrow, Macleod, Buhler, & Loode, 2014).

An important part of the process was supporting project participants to produce "graphic recordings" of their experiences and conversations, at various stages of the program and activities. This strategy is a well-used tool in community activism, since, as Margulies and Sibbet (2007) explain, these methods can help groups communicate and work together. Graphic recording methods:

> have roots in the way designers have always worked, using sketches, diagrams, and imagery to try out new thinking, present possibilities, make sense of complexity, and remember rich amounts of information. Organizations in rapidly changing environments now use visual recorders and graphic facilitators for retreats, planning sessions, team projects, dialogue sessions, problem solving, community building, strategic thinking, and knowledge creation. Educators and trainers use the methods to deepen learning. The applications are extensive and inspiring.
>
> (p. 577)

In research contexts, participatory mapping and visual data collection is often seen as a way to breach generational, language, and professional boundaries (Clark, 2011). These methods also increase the trustworthiness of data interpretation, especially where the investigation is focused on emotional experiences (Copeland & Agosto, 2012). In this project, the graphic recordings produced during the original Brave New Welcomers' creative development week, as well as the recordings generated by the larger youth forum, produced a very evocative visual data set (see Figure 6.3).

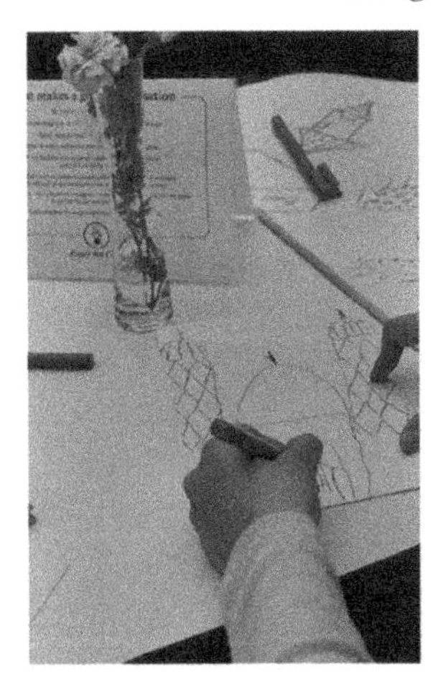

Figure 6.2 The BNW Youth Forum

Figure 6.3 Graphic recordings made by BNW project participants

The Youth Forum formalized our collective intention to extend the gesture of welcome to a wider community. It served as a focal point, a pivotal and performative articulation of three cycles of planning, action, and reflection. In this sense, the forum was a culmination of the synergies between PAR and performed ethnographic methods.

In the end, over a hundred young people from diverse backgrounds participated in the BNW project. All these participants contributed creative content to the project in some form – as digital storytellers; as visual artists; as playback theatre storytellers, devisers, and performers; as co-creative media makers; and as participatory mural makers and graphic recorders. Screening the incomplete BNW film at the Youth Forum was a way of showing our work in progress. Sharing the process of cultural construction in this way works as a mode of research dissemination in the tradition of ethnodrama. Goldstein (2008) explains the technique of performed ethnography as a process whereby ethnographic data is transformed into a script which is, "either read aloud by a group of participants or performed before audiences" (p. 85). The BNW forum itself became a formalised "post-performance conversation," which meant that research participants and audience members were able to have input into the conclusions of the research. According to Goldstein (2008), this can foster more ethical relationships between researchers, research participants, and the communities to which the participants belong (p. 85).

Performing Fieldwork in a PAR Research Project

This BNW project emerged out of work I was involved with for many years – in education, in life story approaches, in socially engaged art, and in action research. It emerged out of the synergies and catalysts sparked by productive partnerships. Out of these beginnings, and over the course of a few years, a study and an activist project drawing on the traditions of PAR and anthropology was established. PAR methods, cycles of planning, action, and reflection (Kemmis, 2009), provided a layered exploration of the ethical challenges involved in working with refugee stories and bringing them into a public sphere. As a social research methodology that provides a structure for blending scholarly, activist, and collaborative objectives, PAR was the central machinery that generated insights and co-created outcomes. Essentially, PAR afforded this program of activist research three main superpowers.

First, PAR was flexible enough to accommodate the "messiness" of this project and its fieldwork. "Mess" is a descriptor often applied to processes and outcomes of action research (Cook, 2009; Goodnough, 2008; Law, 2004), and the BNW project ticked a lot of "messy" boxes. Art by nature is a fluid and messy business, and socially engaged art adds a further element of chaos. Askins and Pain (2011) describe the unpredictability of their research site, referring to a PAR project in the United Kingdom which engaged with young people from diverse cultural backgrounds using arts-based methods. These authors identified a range of ways in which "messiness" manifested in the social space that was activated. The materialities of doing participatory art is, in an obvious sense, messy. Participatory practices are also messy and unpredictable because of the kinds of relationships involved. Further, messiness also is a good descriptor for the politics of interaction, since socially engaged arts can involve fleeting connections and unpredictable encounters, which hold the potential "to cross space, place, and time in unforeseeable ways" (Askins & Pain, 2011, p. 809). Therefore, PAR is an approach that invites us to rise to the challenge of accommodating the unpredictability and complexity of socially engaged arts practice. Kindon, Pain, and Kesby (2007) claim that as an "orientation to inquiry," PAR demands "methodological innovation if it is to adapt and respond to the needs of specific contexts, research questions or problems, and the relationships between researchers and research participants" (p. 13). In this sense, a PAR approach provided the flexibility, tools, and framework to take advantage of the unforeseeable twists and turns, the creative leaps, that emerged as an outcome of the participatory art process. Importantly, while there is always an impulse to impose meaning in the face of complex, dynamic fieldwork, PAR provided the methodological rationale for tolerating "mess" long enough for the research to tell its own story.

The second superpower PAR afforded is the way it supported a focused process of listening, translating, and mediating across various stakeholders' practices and priorities. In other words, PAR is designed to accommodate many voices. In the early stages of scoping this project, as the community engagement process started taking shape as a multipartnership venture with various stakeholders and their expectations of tangible outcomes, PAR met these needs with its explicitly applied approach to research. This project was a complex collaboration, with young participants/storytellers, working in alliance with artists/researchers, activists, peacebuilding, and settlement support professionals. This latter group often worked under the auspices of their employers and were striving to encompass project activities as part of their working lives. Therefore, the input of all these stakeholders was framed by different kinds of parameters. Using PAR meant valuing the needs and inputs of all parties as co-creators of the research and fostering "polyvocality" as well as the "active involvement of research partners in producing knowledge as well as action" (O'Neill, 2011, p. 18). As an outcome of these productive relationships and conversations (Gibson-Graham, 1994), and this considered path through messiness, an overall collective was formed that embarked on a collaborative process of discovery. Using a PAR process gave us the language, right from the start, to acknowledge that what we were planning would involve a trifecta of priorities: action, mutual learning, and co-creativity.

The final superpower PAR afforded was the apparatus for focusing on co-creative practices themselves – on the performance of research, the practice of arts facilitation, settlement support, and peacebuilding. Kemmis (2009) explains that in critical action research, researchers set their sights on transformation, but this also includes transforming the social relationships that support this action. He explains that:

> Thinking of these social formations as 'practice architectures' allows us to think of them as made by people, and thus as changeable by people. People involved in critical action research aim to change their social world collectively, by thinking about it differently, acting differently, and relating to one another differently – by constructing other architectures to enable and constrain their practice in ways that are more sustainable, less unsustainable.
>
> (p. 471)

In this way, a PAR approach allowed us as researchers, arts practitioners, professionals, activists, and as *representatives of the host community,* to focus on the relational aspects of our practice. To focus on the social formations in which the practice occurs, to consider the "practice architecture" of our work and our role as hosts, as things we can construct, reflect on, and change. In short, PAR gave us the methodological tools to orient the research around our own professional and creative practices as well as our roles as members of a host community.

In the end, the Brave New Welcome project became a study in imagined communities, in placemaking and how to *perform* a welcome. Co-creating refugee stories provided a context where a multidisciplinary project team could move out of their practice silos and develop new skills in cultural mediation. These skills allowed us to construct relational spaces where the shape of a good welcome could be explored. We found that creating purposeful, non-denominational spaces, or "contact zones," was essential to the process of generating this welcome. Welcome staged in this way becomes a gesture offered by hosting communities, who are interested in dissolving the boundary between resourced and entitled citizens and newcomers seeking asylum. Within these sites, the facilitators' capacity was critical: the capacity of artists, activists, and educators to collaboratively hold a space occupied by a group of multiple and dissimilar voices long enough for them to explore the texture and nuance of difference, and to develop collective aesthetic and layered expressions of cosmopolitanism. In other words, practitioners in these projects took on the role of cultural translators, who become adept at exploring "social forms of mixture" (Papastergiadis, 2012, p. 142).

Arts-based PAR became a way of mobilising stories to facilitate a "promiscuous traffic between different ways of knowing" (Conquergood, 2002, p. 145). In this sense, the work of the facilitators became a praxis of cultural translation and story stewardship, the effectiveness of which hinges on the materiality of the practice. Storytelling, expressed in an aesthetic form via visual art, theatre, and participatory media, is a fluid and potent form of communication, which nonetheless has an inescapable materiality. This is captured in Conquergood's (2002) evocation of French philosopher Michel de Certeau's pithy phrase "what the map cuts up, the story cuts across" (p. 145). This analogy describes the "transgressive travel between two different domains of knowledge: one official, objective, and abstract – "the map"; the other one practical, embodied, and popular – 'the story'" (Conquergood, 2002, p. 145). If we understand that refugee storytelling has the potential to produce imaginary spaces that transcend local and national geographical boundaries, then we can then begin to work on how to put this imaginary to use in practical ways.

Methodological Innovation

In the process of working out how to enact this kind of welcome, we discovered that performance studies offered another strand of theory and practice useful to achieve the kind of "methodological innovation" we needed. The BNW Project brought PAR approaches into dialogue with a form of critical and performed ethnography. Conquergood's body of work (1985, 1988, 1991, 2002) provided a useful historical reference for the hybrid nature of our activist research, as well as his articulation of critical performance ethnography as a research approach founded in dialogue, praxis, and a politics of resistance. Conquergood brought two important ideas together in a way that has resonated in performances studies for several decades. He wrote about a research approach that supports the idea of inquiry as creative intervention. He also identified performance as a site of resistance and interruption. Conquergood's performance studies agenda arose from extensive fieldwork, including performance projects with Hmong refugees, Laotian refugees in Thailand, and Palestinian refugees in the Gaza Strip. Performance studies as he conceived and performed it represents an unravelling of the textual tethering that has been the historical focus of ethnography and much of cultural studies. He spoke of "an ethnography of the ears and heart that reimagines participant-observation as co-performative witnessing" (2002, p. 142). Conquergood (2002) locates the researcher as an active agent in a "experiential participatory epistemology" (p. 149), which embraces three different interwoven ways of knowing. Such a stance involves bringing into dialogue what he calls, "the three a's of performance studies: artistry, analysis, activism" (Conquergood, 2002, p. 152). It involves forging "a unique and unifying mission around the triangulations of these three pivot points" (Conquergood, 2002, p. 152). Elaborating on this alliterative construct, Conquergood (2002) explains these pivot points as:

> 1. Accomplishment – the making of art and remaking of culture; creativity; embodiment; artistic process and form; knowledge that comes from doing, participatory understanding, practical consciousness, performing as a way of knowing.
>
> 2. Analysis – the interpretation of art and culture; critical reflection; thinking about, through, and with performance; performance as a lens that illuminates the constructed creative, contingent, collaborative dimensions of human communication; knowledge that comes from contemplation and comparison; concentrated attention and contextualization as a way of knowing.
>
> 3. Articulation – activism, outreach, connection to community; applications and interventions; action research; projects that reach outside the academy and are rooted in an ethic of reciprocity and exchange; knowledge that is tested by practice within a community; social commitment, collaboration, and contribution/intervention as a way of knowing: praxis.
>
> (p. 152)

The catalyst and the lifeblood for this study were the life stories that young people from refugee backgrounds were willing to share. It was clear from the start, however, that research engaging with these stories and these storytellers would need to be based on an inventive and collaborative form of communication. In homage to Conquergood's ethics of research practice, it would also need to be reciprocal in its operations. The ambitious plan that was being shaped was to form alliances and forge a program of work anchored by the triangulation of the three pivot points Conquergood (2002) wove together with such poetry – artistry, analysis, activism. Richard Schechner (2013) frames this movement between "sympathetic participation" in a performance project, and the adoption of a position of critical distance, as a "Brechtian" kind of approach to fieldwork. German poet, playwright, and theatre director Bertolt Brecht was a Marxist who made highly political theatre, known as "epic" or "dialectic theatre" until his death in 1956. The theatrical techniques he is famous for were designed to eschew escapism and expose his audiences to political realities. The "distancing," "estrangement," or "alienation" effect prevented the audience from losing itself completely in the narrative. A range of theatrical strategies were used to encourage audiences to look at social issues critically, such as consciously drawing attention to the craft of theatre making itself, and confronting audiences with uncomfortable provocations.

Performance studies, according to Schechner (2013, p.2), appropriates "participant observation" (a highly valued method adapted from anthropology) and reinvents it. In anthropological fieldwork, participant observation is a way of learning about cultures other than that of the fieldworker. In anthropology, for the most part, the "home culture" is Western and the "other" non-Western. But in performance studies, the other may be a part of one's own culture (non-Western or Western), or even an aspect of one's own behaviour. That positions the performance studies fieldworker at a Brechtian distance, allowing for criticism, irony, and personal commentary, as well as sympathetic participation. In this active way, as Schechner explains, one can "perform" fieldwork.

It is in relation to the creative outputs that the PAR approach was brought into dialogue with a form of critical and performed ethnography. This approach can be located within a trajectory of anthropology developed out of the "crisis of representation," or what critical ethnographer Soyini Madison (2012) explains has been variously identified as "the performance turn, 'the postmodern turn,' or the 'new ethnography'" (p. 13). Lassiter and Campbell (2010) explain how critical ethnography, as a new or reinvented approach to research can be seen as:

> coming of age in the wake of the 1980s critique, (and engaging us) in cooperative approaches to research that imagine and push toward, in deliberate and explicit

ways, coinscription, corepresentation, and, in turn, collaborative actions: all connected as a continuum, or constellation, of praxis.

(p. 758)

Barbara Tedlock (2007), speaking of the same movement toward explicitly collaborative, activist, and performative approaches to research, explains that ethnographers are engaging with forms that provide an interface with the public, in an effort to respond to these new mandates. Forms such as experimental theatre, personal narratives, filmmaking, and documentary photography expose the mechanics of representation. These forms "produce mimetic parallels through which the subjective is made present and available to its performers and witnesses" (Tedlock, 2007, p. 160).

It was not until fieldwork was well underway that I began to see how the PAR framework could be complemented by this strand of public ethnography. The creative works produced in these projects, and the public events where they were shared and invited interaction with audience members, were forms of research dissemination. Lassiter (2008) explains that public ethnography "endeavours to engender texts that are more readable, relevant, and applicable to local communities of ethnographic collaborators (i.e., local publics)" (p. 73). According to Tara Goldstein (2008), performed ethnography involves "turning educational ethnographic data and texts into scripts and dramas that are either read aloud by a group of participants or performed before audiences" (p. 85). Tedlock (2007) positions this tradition of ethnographically derived performance within a genre of political theatre. She describes an emblematic example of "ethnodrama" involving an alliance between an anthropologist/dramaturg and a popular theatre company where cast and audience members engage in a dialogue at the conclusion of a performance. Here, in the tradition of a Brechtian style of "epic theatre" (consciously provoking audiences to engage in dialogue with actors and theatre makers), "ideas for ways to improve the production as a work of art, cultural document, and political critique are aired, and changes are included in future performances" (Tedlock, 2007, p. 156).

Goldstein (2008) claims that this kind of playwriting and production have powerful applications in educational research contexts, since they "both shape and show cultural construction in action" (p. 85), and refine representation via a reiterative or "closed loop" approach. Goldstein outlines a process whereby post performance conversations are held, involving research participants and audience members who are then are to input into the conclusions of the research. According to Goldstein (2008):

> The incorporation of audience input into on-going revisions of the play provides an opportunity for mutual analysis, and in doing so, can help create more ethical relationships between researchers, their research participants, and the communities to which the research participants belong. Post-reading/performance conversations also allow ethnographers in education to link up their research to their teaching and larger public forums on pressing social issues.
>
> (p. 85)

In a contemporary research context, expanded, hybridised forms of digital/performed ethnography, drawing on the traditions of epic theatre, are finding a home in a variety of contexts: social policy (Sandercock & Attili, 2010), art activism (Sarkar, 2013), radical pedagogy (Harris, 2010), and at the nexus of globalism and education (Gallagher & Wessels, 2013; Puwar, 2012), among others. As a creative collaboration, the

BNW project followed the contours of traditional ethnodrama in some respects, and experimented with the form, like the examples mentioned above. For example, the scripts for the film and the visual elements were carefully and collaboratively crafted as research outcomes; these were artistic expressions and representations of ethnographic data. The film script is the result of what anthropologist Victor Turner calls "plural" or "collective reflexivity" (1982, p. 75–76). The collaborative process of editing the BNW film constituted a form of collective thematic coding. The film was supported by performances, forums, and participatory art projects, events which were consciously staged in public contexts in a manner that invited audience feedback and interaction. The BNW film was screened as a work in progress. It was presented to a public audience and the feedback and input from participants at the BNW Youth Forum were incorporated into the script for the final film.

One of the benefits of this new understanding was that I now had a way to make sense of the tensions and conflicts this multimodal form of research produced, and to account for the various outcomes. The whole team of practitioners experienced these conflicts. Goldstein (2008) writes that, "as a writing method that links ethnographic data analysis to dramatic writing, dramatic performance to critical conversation and discussions, performed ethnography demands multiple commitments of the researcher, which sometimes compete and lie in tension with one another" (p. 89). As a team of practitioners, we were aware of these multiple demands and the ongoing ethical challenges this tension produced. Goldstein (2008) explains that:

> While all research has multiple audiences to be accountable to, the hybrid form of performed ethnography – part ethnography, part drama – requires the researcher-playwright to satisfy the social science demands of ethnography and the aesthetic demands of drama. When performed ethnography is also linked to goals of civic engagement and social change, there are pedagogical and dialogical demands to satisfy as well.
>
> (p. 98)

Despite this underlying (and ultimately unresolvable) tension, the creative works generated by the BNW project served as logical and empowering outcomes of our PAR and activist approach. They served as potent methods for engaging audiences, functioning as effective "strategies for staging interventions" (Conquergood, 2002, p. 151). In addition, as Goldstein (2008) argues, the performative and dialogic elements provided the work, "with 'internal' (Lincoln & Guba, 1985) or 'face' (Lather, 1986) validity, which is important in discussions of rigor in ethnography" (p. 88).

As an extension of traditional ethnodramatic approaches, the BNW project integrated digital and performative elements. According to Gallagher et al. (2013), this kind of hybridity can enhance the options for validity and member checking. Gallagher et al.'s (2013) multi-sited ethnography engaged with school sites in Canada, India, Taiwan, and the United States to investigate the experiences of young people "often marked as 'disadvantaged' and 'marginal' to the traditional practices of schooling" (p. 177). An important discovery in their work was that the hybridity of live, digital, and textual means of communication and expression, woven into an ethnographic engagement with global youth, has powerful potential to address questions of power and voice. Live and digital methods combined to support new and enriched understandings of the experiences and perspectives of young people. Gallagher et al. (2013) note:

> Unlike a 'member check,' however sophisticated, after the research report has been written, digital tools, like video-making and blogging, play with authorship in the ethnographic field and invite participant analysis in situ rather than post facto . . . We have argued here, with grounded empirical evidence, that the very hybridity of the live and the digital represents, in our view, a cosmic shift, as to how ethnographic research on/with/for young people might be transformed in this next moment of ethnography.
>
> (p. 191)

Ultimately, I came to see that the collaboratively produced creative works functioned in the manner that Gallagher et al. (2013) describe – they worked to "play with authorship in the ethnographic field." These creative works formed parallel texts – expressions that, as Conquergood (2002) claims, can "stand alongside and in metonymic tension with" the written narrative/report of the project and my data analysis (p. 151). In this way, the original intention to focus the study on "hosts" rather than "newcomers" as research subjects could be honoured. The co-created art and media became the carrier of young people's voice – communication that was authored directly by this group rather than being translated by the researcher's accounts and analysis. In this study, the co-created art and media served as both the research method as well as the findings. To put it more axiomatically, these creative works are both the medium and the message.

Listening to the Listeners

Digital Storytelling projects and oral history projects often come about because a community of people find they have common ground in some way, they can work well on several levels when there is a larger narrative, a metanarrative that frames a storytelling project (Lénárt-Cheng & Walker, 2011; Thumim, 2009). This may be a collection of stories about an experience of homelessness, for example, or about surviving mental illness, or about belonging to a place. Quite often the listeners and storytelling facilitators in these projects are those who have a place in the storytelling community.

In the BNW project, however, the relationship was determined by political reality and colonial history, as well as by a relationship of difference between the roles of host and newcomer, at least initially. Although this dualism started to break down quickly, once the groups of storytellers and project participants grew and became more diverse, it was the elephant in the room during the generation of the project and the first few workshops. Differences in migration, citizenship status, and language between those who spoke English as a first language (the team of project organisers and facilitators and Australian born young people) and those who were learning it as an additional language, were ascendant and undeniable. Being a listener, a storytelling mediator, therefore required that we all moved to new territory. It required that we create what Homi Bhabha (1994) called a "third space" and what Mary Louise Pratt (1991) called a "contact zone."

As both PAR and public ethnography are methodologically and epistemologically grounded in an analysis of power dynamics, the wider political and sociological implications of intercultural communication were constantly brought into focus. In this sense, the notion of "contact zones" became a useful vector. The term "contact zone" comes from literary and critical theorist Mary Louise Pratt (1991). Pratt (1991) defined contact zones as the social spaces where cultures meet, clash, and grapple with each other, often in contexts of highly asymmetrical relations of power, such as colonialism, slavery, or their aftermaths as they are lived out in many parts of the world today" (p. 34). The concept has been taken up in a range of disciplines where postcolonial, transcultural, inclusionist, and collaborative practices are valued, including: youth participatory action research (Askins & Pain, 2011; Torre et al., 2008), postcolonial approaches to museum exhibition design (Clifford, 1997; Hutchison & Collins, 2009), and in anti-racist education (Wolff, 2002), just to name a few examples and contexts. Askins and Pain (2011) claim that in intercultural research it is particularly valuable to consider 'contact zones' as somewhat uncomfortable spaces, characterised by productive tensions, since this foregrounds issues of privilege, and "necessitates working with and through issues of voice, agency, power, and desire alongside all participants in the process" (p. 807).

At the beginning of the fieldwork, as new partnerships were being formed, I was intent on finding a way to structure research space that invited, "a textured understanding of human interaction across power differences" (Torre et al., 2008, p. 25). My solution to this task, although never finalised or perfected, centred on the idea of "listening to the listeners" (Jeffers, 2011). The construct of a contact zone was helpful in the process of defining the layers of listening involved in this study, what I came to describe (after Alison Jeffers, 2011) as a process of "double translation." As the storytelling activities evolved, I began to make connections between the questions framing the study and the decisions I was grappling with about the research design. Jeffers (2011), a British scholar of refugee theatre, writes:

> Interpreting refugees' stories for a western audience involves a process of translation; as a scholar of refugee theatre, the process of listening in performance requires a kind of double translation. . . . It is important to listen to the listeners – the writers, actors and directors who create theatre and performance works concerning refugees, while maintaining the imperative to listen to refugees themselves.
>
> (p. 2)

As the fieldwork began in earnest, I observed that as a researcher in this context I was exploring a similar layered positionality that Jeffers (2011) referred to when she considered her role. I concluded that the PAR project I was engaged in was one that required a double translation, a dual listening that she described. I had already realized that the stories young people from refugee backgrounds were sharing were central to the whole project. This meant I cast myself firstly as a listener of these stories, someone who was working with my collaborators to enable the telling of these stories, and there were layers of complexity involved in taking this listening position. The task of communicating, of listening, across cultures was an obvious challenge, but even before any stories were shared, the listening began with the process of building a third space, a contact zone where relationships based on trust could be negotiated.

The creative works produced by the BNW project arose in the intersection of the collective of practitioners and participants – an intercultural contact zone. Working in this zone, we focused on using participatory art, visual art, drama, and co-creative media, and an aesthetic mode of expression. The tangible artistic outcomes emerging from this contact zone consist of the performances, the short film, the collaborative artwork, and graphic recordings. As part of the research, these products were shared with wider local audiences, screened and staged at separate events. Participatory artworks were generated at the forum and at other public screenings. These artefacts were also uploaded to online platforms such as Vimeo (the film) and Facebook (the film, graphic recordings, and photographs). These creative outcomes needed no extra translation. They exist because of the shared creative language that was generated among the participants, with the support of the facilitating team.

In addition to the process of listening described above, this study also focused on "listening to the listeners" – my colleagues. My task was to analyse and theorise about this process of creating a contact zone where stories can be shared, of working collaboratively with refugees. My task was to understand more about how settlement support workers, advocates, arts workers, technologists, filmmakers, cultural workers, and so on do their work; to render those stories into an aesthetic and digitalised form; to understand how we translate and frame refugee stories for a mainstream western audience; and how we generate events and platforms for story sharing, manufacturing audiences, and publics. This was the second layer of listening, the second layer of translating.

What we learnt as hosts and cultural practitioners was that common ground cannot be assumed; it must be consciously created from the ground up. The BNW project affirmed for us that PAR and arts-based interventions are especially effective models for this kind of work. PAR directs us to carefully consider issues of power relations inherent in a research encounter. In this storytelling project, a participatory arts process allowed us as facilitators to build an aesthetic milieu for sharing stories, one that worked as affect and metaphor so that a new vocabulary, a new grammar, and new ways of knowing could be expressed and shared. Public ethnography provides tools for ethically scaling up stories and manufacturing public storytelling spaces. Woven together, these principles and methods formed a model for building a third space. This model recognises that such spaces are the result of generating contact zones, which can only happen when social relationships are sustained for long enough to build understanding across difference.

Importantly, creating a working contact zone was in turn a critical foundation for the next stage of orchestrating a public space for sharing the stories. Cultural philosopher Arjun Appadurai's (2002) views on the role of the imagination underscore the need to keep these public storytelling spaces alive, because they are critical in the process of infusing meaning into urban space. Appadurai (2002) evokes the ongoing nature of this project and prompts another critical question:

> The city is where we have to do some trial and error. I don't have a formula for a solution, except for one thing. Let's, in seeking solutions endorse the following principle. Even the poorest of the poor should have that capability, the privilege and the ability to participate in the work of the imagination. Can we create a politics that recognises that? That is the question.
>
> (p. 46)

The work of storytelling facilitators in this context is to function as activists who recognise the rights of refugees and asylum seekers to "participate in the work of the imagination" (Appadurai, 2002, p. 46). In this sense, we work as co-constructers of an aesthetic milieu and as wielders of the "geographies of matter" (Askins & Pain, 2011) in the service of intercultural communication. Our work as interpretive framers, as cultural translators, and as mediators of a cosmopolitan sensibility, function to make this possible. For sociopolitical and relational understandings to be set in motion when we hear these stories, so they make sense in a teleospatial way. A spatial sensibility, what Doreen Massey (2005) calls a "Russian-doll ethical imaginary" (p. 187), dictates the degree of our ethical engagement with others, so that we feel a stronger ethical obligation to those closest to us. Refugee stories staged in a public space make a relational claim in a civic space that disrupts the placement of refugees and asylum seekers at the extreme edge of our frame. Stories can make links between newcomers and hosts, and thereby link the new arrival to the city itself. Once a newcomer is recognised through their story, a new cultural imaginary can take shape, one that sends offshoots out from the cluster of interconnected stories, so that the city itself is a knot of stories (Ingold, 2008), rather than an enclosed place with boundaries and borders to be guarded.

Walking the Fine Line Between Pedagogy and Art

Operationalising storytelling in the service of conducting participatory research, and "teaching" the public something about refugee lives presumes instrumental goals. There was a constant sense in these projects that as practitioners we were walking a fine line between action, pedagogy, and art. The tension between instrumental and aesthetic goals in creative practice has received much critical attention. Some scholars argue that an instrumental approach to evaluating creative practice is misplaced. Since this project wanted to produce a film as part of a strategy to shift perceptions and attitudes, and to influence and inspire audiences to become part of the Brave New Welcome movement, we did not completely sidestep instrumental goals. We were concerned with the wellbeing of the participants and the value placed on authentic participatory process. Nevertheless, we needed "good" art, something that would affect audiences.

In reviewing the role of community and public art in the era of neoliberalism, Papastergiadis (2012) observes that the "discourse on the political significance of art is still trapped in the debate over whether or not it can make a difference in the overall social context" (p. 174). He argues that the result of this misconception is a quality of critical attention which misses the point of collaborative art practice, since:

> The effects of art tend to be registered only to the extent that they appear outside of its own, apparently autonomous, field. Is art only a value when it transforms or reflects the social? This question presumes that art is external to the existing forms of the social and must be something to the social in order to have a viable function. The place and function of art, as always, operates within the social. However, the new collaborative movements have sought to take an active role in social change, not by means of radical intervention for critical reflection, but through the mediation of new forms of cosmopolitan knowledge.
>
> (p. 175)

Mediating new forms of sharing stories with refugees and asylum seekers became, almost by default, an exercise in mediating this new form of cosmopolitan knowledge that Papastergiadis discusses. As practitioners and cultural agents operationalising refugee storytelling to enact a morally engaged form of hospitality, we were called upon to consciously balance our roles of translation, co-creation, education, and peacebuilding. We worked hard to rise to the challenge of performing as civic minded activists, storytelling champions, and cultural mediators. At our most hopeful moments, we found that this kind of dynamic integration of socially engaged, aesthetic, and activist priorities – or "co-performative refugee storytelling" – can shift the interaction between tellers and listeners. It shifts out from underneath the regime of fear, pity, and rescue, to recast the storytellers as co-producers of culture in a new place. Amid the intractable politics of forced migration and border control, co-performing refugee stories can be a practice that claims authentic engagement, a practice that supports us to rehearse a cosmopolitan sensibility (Papastergiadis, 2012).

Conclusion

The collective starting point that drove the development of the BNW storytelling project was to explore using personal stories to invite audiences to orchestrate a better welcome for young people from refugee backgrounds. We wanted to invite our fellow city dwellers to step over the boundary of their imagined community of insiders and outsiders, to recognise a responsibility to respond to the human cost and global fallout of armed conflict, even though they might not share these experiences first-hand. Gillian Whitlock (2007) characterises life narratives employed in this fashion as "soft weapons." The aim of this project was to harness the power personal narratives can wield to invoke a moral response among Brisbane residents. We wanted these publics to consider more carefully their own role as hosts in a place of refuge and to join the struggle for change. We want them to care. We wanted to build solidarity. Ultimately, audience engagement in the scaled up public spaces we created meant we managed to achieve those aims, at least to some degree.

While the stories that emerged served those purposes well, they also had a life of their own. In the end, the project became an extended act of cultural mediation with microcosmic beginnings. The project became an act of performative translation that gathered meaning and momentum and scope. It became a project that was different things to different people. In this chapter, I discussed the (inevitably imperfect) resolutions I arrived at as a practitioner-scholar to address issues of voice and representation, and to accommodate the messiness of knowledge that emerges from participatory arts projects within research. I describe how using PAR, public ethnography, and performative inquiry methods made it possible to extend an aesthetic and playful antenna into the problems and questions framing the study. They made it possible to highlight new understandings that emerged at the interface of different modes of action, experimentation, and knowledge production, and made it possible to frame fieldwork as a co-performative practice. In this chapter, I traced the way our efforts to scale up the refugee stories generated by this project meant that we all learned something imagined and something visceral. We learned how to embrace the complex task of promoting diversity and inclusion in urban spaces in ways that are more genuinely open and more genuinely brave.

Acknowledgements

The author gratefully acknowledges funding support which made the fieldwork reported here possible, from the Australian Research Council for the Linkage Project, "Digital Storytelling and Co-creative Media: The role of community arts and media in propagating and coordinating population wide creative practice" (LP110100127). The author also acknowledges the contributions of other members of the facilitating team in the concept, design, and implementation of the project. This co-creative team included Madeleine Belfrage, Cymbeline Buhler, Angus Macleod, Erica Rose Jeffrey, and Serge Loode, along with the storytellers themselves: Machemeh, Waniya, Daniel, Solly, Andres, Asad, Jacob, Laura, Aref, Brea, Arwin, Nadir, Darwood, Fiarrah, Rosie, Zabi, Sarah, Erin, and Concy.

References

Appadurai, A. (2002). The right to participate in the work of the imagination. Interview by Arjen Mulder. In J. Brouwer, P. Brookman, & A. Mulder (Eds.), *TransUrbanism*. 33–47. Rotterdam: NAi Publishers.

Askins, K., & Pain, R. (2011). Contact zones: Participation, materiality, and the messiness of interaction. *Environment and Planning-Part D*, 29(5): 803–821. https://doi.org/10.1068%2Fd11109.

Bhabha, H. (1994). *The Location of Culture*. New York: Routledge.

Brown, J., Homer, K., & Isaacs, D. (2007). The world cafe. In P. Holman, T. Devane, & S. Cady (Eds.), *The Change Handbook: The Definitive Resource on Today's Best Methods for Engaging Whole Systems*. 326–361. San Francisco: Berrett-Koehler Publishers.

Clark, A. (2011). Multimodal map making with young children: Exploring ethnographic and participatory methods. *Qualitative Research*, 11(3): 311–330. https://doi.org/10.1177%2F1468794111400532.

Clifford, J. (1997). Museums as contact zones. In J. Clifford (Ed.), *Routes: Travel and Translation in the Late Twentieth Century*. 188–219. Cambridge, MA: Harvard University Press.

Conquergood, D. (1985). Performing as a moral act: Ethical dimensions of the ethnography of performance. *Literature in Performance*, 5(2): 1–13. https://doi.org/10.1080/10462938509391578.

Conquergood, D. (1988). Health theatre in a Hmong refugee camp: Performance, communication, and culture. *TDR*, 32(3): 174–208. https://doi.org/10.2307/1145914.

Conquergood, D. (1991). Rethinking ethnography: Towards a critical cultural politics. *Communication Monographs*, 58(2): 179–194. https://doi.org/10.1080/03637759109376222.

Conquergood, D. (2002). Performance studies: Interventions and radical research. *TDR/The Drama Review*, 46(2): 145–156. https://doi.org/10.1162/105420402320980550.

Cook, T. (2009). The purpose of mess in action research: Building rigour though a messy turn. *Educational Action Research*, 17(2): 277–291. https://doi.org/10.1080/09650790902914241.

Copeland, A. J., & Agosto, D. E. (2012). Diagrams and relational maps: The use of graphic elicitation techniques with interviewing for data collection, analysis, and display. *International Journal of Qualitative Methods*, 11(5): 513–533. https://doi.org/10.1177%2F160940691201100501.

Fox, H. (2007). Playback theatre: Inciting dialogue and building community through personal story. *The Drama Review*, 51(4): 89–105. https://doi.org/10.1162/dram.2007.51.4.89.

Gallagher, K., & Wessels, A. (2013). Between the frames: Youth spectatorship and theatre as curated, "unruly" pedagogical space. *Research in Drama Education: The Journal of Applied Theatre and Performance*, 18(1): 25–43. https://doi.org/10.1080/13569783.2012.756167.

Gallagher, K., Wessels, A., & Ntelioglou, B. Y. (2013). Becoming a networked public: Digital ethnography, youth and global research collectives. *Ethnography and Education*, 8(2): 177–193. https://doi.org/10.1080/17457823.2013.792507.

Gibson-Graham, J. K. (1994). "Stuffed if I know!": Reflections on post-modern feminist social research. *Gender, Place and Culture: A Journal of Feminist Geography*, 1(2): 205–224. https://doi.org/10.1080/09663699408721210.

Goldstein, T. (2008). Performed ethnography and the pursuit of rigor. In K. Gallagher (Ed.), *The Methodological Dilemma: Creative, Critical and Collaborative Approaches to Qualitative Research*. 85–102. New York: Routledge.

Goodnough, K. (2008). Dealing with messiness and uncertainty in practitioner research: The nature of participatory action research. *Canadian Journal of Education*, 31(2): 431–458. https://journals.sfu.ca/cje.

Harris, A. (2010). Race and refugeity: Ethnocinema as radical pedagogy. *Qualitative Inquiry*, 16(9): 768–777. https://doi.org/10.1177%2F1077800410374445.

Hume-Cook, G., Curtis, T., Woods, K., Potaka, J., Tangaroa-Wagner, A., & Kindon, S. (2007). Uniting people with place using participatory video in Aotearoa New Zealand: a Ngāti Hauiti journey. In S. Kindon, R. Pain, & M. Kesby (Eds.), *Participatory Action Research Approaches and Methods: Connecting People, Participation and Place*. 186–195. New York: Routledge.

Hutchison, M., & Collins, L. (2009). Translations: Experiments in dialogic representation of cultural diversity in three museum sound installations. *Museum and Society*, 7(2): 92–109. https://journals.le.ac.uk/ojs1/index.php/mas/index.

Ingold, T. (2008). Bindings against boundaries: Entanglements of life in an open world. *Environment and Planning A: Economy and Space*, 40(8): 1796–1810. https://doi.org/10.1068%2Fa40156.

Jeffers, A. (2011). *Refugees, Theatre and Crisis: Performing Global Identities*. London: Palgrave Macmillan.

Kemmis, S. (2009). Action research as a practice-based practice. *Educational Action Research*, 17 (3): 463–474. https://doi.org/10.1080/09650790903093284.

Kindon, S., Pain, R., & Kesby, M. (2007). Introduction: Connecting people, participation and place. In S. Kindon, R. Pain, & M. Kesby (Eds.), *Participatory Action Research Approaches and Methods: Connecting People, Participation and Place*. 1–6. New York: Routledge.

Kunt, Z. (2020). Art-based methods for Participatory Action Research (PAR). *Interactions: Studies in Communication & Culture*, 11(1): 87–96. https://doi.org/10.1386/iscc_00008_1.

Lassiter, L. E. (2008). Moving past public anthropology and doing collaborative research. *Napa Bulletin*, 29(1): 70–86. https://doi.org/10.1111/j.1556-4797.2008.00006.x.

Lassiter, L. E., & Campbell, E. (2010). What will we have ethnography do? *Qualitative Inquiry*, 16(9): 757–767. https://doi.org/10.1177%2F1077800410374444.

Lather, P. (1986). Issues of validity in openly ideological research: Between a rock and a soft place. *Interchange*, 17(4): 63–84. https://doi.org/10.1007/BF01807017.

Law, J. (2004). *After Method: Mess in Social Science Research*. London: Routledge.

Lénárt-Cheng, H., & Walker, D. (2011). Recent trends in using life stories for social and political activism. *Biography*, 34(1): 141–179. https://www.jstor.org/stable/23541185.

Lincoln, Y. S., & Guba, E. G. (1985). *Naturalistic Inquiry*. Thousand Oaks, CA: Sage.

Madison, D. S. (2012). *Critical Ethnography: Method, Ethics, and Performance*. Thousand Oaks, CA: Sage.

Margulies, N., & Sibbet, D. (2007). Visual recording and graphic facilitation: Helping people see what they mean. In P. Holman, T. Devane, & S. Cady (Eds.), *The Change Handbook: The Definitive Resource on Today's Best Methods for Engaging Whole Systems*. 366–341. San Francisco: Berrett-Koehler Publishers.

Massey, D. (2005). *For Space*. Thousand Oaks, CA: Sage.

O'Neill, M. (2011). Participatory methods and critical models: Arts, migration and diaspora. *Crossings: Journal of Migration & Culture*, 2(1): 13–37. https://doi.org/10.1386/cjmc.2.13_1.

Papastergiadis, N. (2012). *Cosmopolitanism and Culture*. Cambridge: Polity Press.

Pratt, M. L. (1991). Arts of the contact zone. *Profession*, 33–40. https://www.jstor.org/stable/25595469.

Puwar, N. (2012). Mediations on making Aaj Kaal. *Feminist Review*, 100(1): 124–141. https://doi.org/10.1057%2Ffr.2011.70.

Raynor, R. (2019). Speaking, feeling, mattering; Theatre as method and model for practice-based, collaborative, research. *Progress in Human Geography*, 43(4): 691–710. https://doi.org/10.1177/0309132518783267.

Sandercock, L., & Attili, G. (2010). Digital ethnography as planning praxis: An experiment with film as social research, community engagement and policy dialogue. *Planning Theory & Practice*, 11(1): 23–45. https://doi.org/10.1080/14649350903538012.

Sarkar, S. (2013). Arts, activism, ethnography: Catapult Arts Caravan, 2004–2010. *Museum Anthropology Review*, 7(1–2): 155–165. https://scholarworks.iu.edu/journals/index.php/mar/index.

Schechner, R. (2013). *Performance Studies: An Introduction*. New York: Routledge.

Tedlock, B. (2007). The observation of participation and the emergence of public ethnography. In N. K. Denzin & Y. S. Lincoln (Eds.), *The SAGE Handbook of Qualitative Research*. 151–172. Thousand Oaks, CA: Sage.

Thumim, N. (2006). Mediated self-representations: 'Ordinary people' in 'communities'. In S. Herbrechter & M. Higgins, *Returning (to) Communities*. 255–274. Amsterdam: Rodopi. https://doi.org/10.1163/9789004325623_016.

Tolia-Kelly, D. (2007). Participatory art: Capturing spatial vocabularies in a collaborative visual methodology with Melanie Carvahlo and South Asian women in London, UK. In S. Kindon, R. Pain, & M. Kesby (Eds.), *Participatory Action Research Approaches and Methods: Connecting People, Participation and Place*. 691–710. New York: Routledge.

Torre, M. E., Fine, M., Alexander, N., Billups, A. B., Blanding, Y., Genao, E., et al. (2008). Participatory action research in the contact zone. In J. Cammarota & M. Fine (Eds.), *Revolutionizing Education: Youth Participatory Action Research in Motion*. 23–44. New York: Routledge.

Turner, V. W. (1982). *From Ritual to Theatre: The Human Seriousness of Play*. New York: Performing Arts Journal Publications.

Whitlock, G. (2007). *Soft Weapons: Autobiography in Transit*. Chicago: University of Chicago Press.

Wolff, J. M. (2002). *Professing in the Contact Zone: Bringing Theory and Practice Together*. Urbana, IL: National Council of Teachers of English.

Woodrow, N., Macleod, A., Buhler, C., & Loode, S. (2014). *Brave New Welcome*. Vimeo. https://vimeo.com/98102090.

7 "You Think Too Much!"

Emotional Geographies of Participatory Action Research

Kye Askins

Let me begin by recalling two moments of insight during Participatory Action Research (PAR) fieldwork. The first came towards the end of a project among participants brought together through a befriending scheme. We were holding a workshop to check in about emerging findings regarding their encounters as refugees, asylum seekers, and more settled residents. I was feeding back on how we had identified that spending time face-to-face is important to people's shifting understandings of others and themselves, and to building social capacities and community cohesion. I was trying to explain in everyday language how *becoming* is a critical feminist concept that challenges binary constructions and understandings of fixed identities and positions. Rather than talking about *being* together, I suggested we discuss *becoming* together, at which point one woman laughed, slapped her hand down on the table and said:

> Kye, you think too much! Being together is enough, it's because we spend time together that we care about each other, that's what's important [much nodding and agreement around the room].
>
> (personal fieldnotes, February 2014)

This gentle rebuff further validated the key theme above: that physical proximity/presence is vital to intercultural understanding and building bonds, and to people 'recognising *with*' each other (Noble, 2009). Through complicated, *embodied, and emotional* entanglements facilitate connection through caring. It also told me not to project an overly academic interpretation onto participants.

A second moment is equally central to the theme of this chapter:

> A few of us were sitting round the table this morning, planning general Drop-In and specific research activities. John (man, white, Scottish, 50s, long-term resident[1]) had just had a very short haircut, which several had commented on already. Adam (man, black, Sudanese, 40s, claiming asylum) came in late, said hi, then went round behind John and "ruffled" his hair. John and everyone laughed, Adam grinned, saying "where's it all gone?!"
>
> I was amazed! John's a pretty tough-sounding Glaswegian, whose appearance doesn't invite anyone to get close, let alone to run their hands over his head. This moment of intimacy between two men, both from cultures of strong masculinity/patriarchy, really struck me. Yet everyone else around the table seemed to take it in their stride. I asked Helen (woman, white, Australian, 40s, long-term

DOI: 10.4324/9780429400346-7

resident) later if she was surprised by their closeness and tactility, and she said "Och no, we wear our hearts on our sleeves here!"[2]

(personal fieldnotes, April 2016)

This exchange deeply moved me, affecting my relationships with participants, and encouraging me to also "wear my heart on my sleeve" as a volunteer-researcher.[3] These moments, and many more, have taught me that emotional intelligence and openness in research, rather than hiding or neglecting emotions, is key to PAR – to elicit, enable, and understand the voices and stories of people, and the relations between them.

This chapter draws inspiration from 18 years of PAR research in England and Scotland. It's a very personal and intersubjective attempt to demonstrate how rationally considered accounts (from both participants and academic researchers) are simultaneously embedded in difficult power relations, structural violences, identity politics, and sensuous, embodied, and emotional entanglements.

Following feminist, decolonial, and other critical scholarship, I seek to provide a tender, cautious, critical mapping of the emotional geographies of PAR to "enflesh and decolonize representation [and] nurture alternative ways of thinking, being, doing and loving" (Motta & Seppala, 2016, p. 7). I'm concerned with how holding onto emotions, without losing the capacity for critical, rational thought, enables research to nurture conditions of potential with others, "queering boundaries of academic knowledge production" and queering dominant white, masculinist forms of "knowing" that devalue and ignore feelings and senses (Motta & Seppala, 2016, p. 7). While there is increasing research on emotions in space and society (Davidson, Bondi, & Smith, 2015), this chapter is agitated with *working with and through the emotional geographies of doing* PAR – whatever the topic, issue, research questions or groups participating. In this, I am heedful that:

"You Think Too Much!"

Being told that I think too much really got under my skin and niggled away in my brain. While this statement articulated the centrality of feelings in society and space (something I have long researched), it too easily dismissed the rational part of socio-political living that is also valid. I knew-felt[4] a dissonance, since throughout the research process it was evident that participants thought about and reflected on their roles and relationships very much.

My discomfort illustrates one of PAR's central challenges: how to enact, understand, and represent processes of co-production. A key aim of PAR is "to enable spaces for collaboration, negotiation and the co-construction of knowledge" (Wynne-Jones, North, & Routledge, 2015, p. 218). Therefore, when writing up research, it is imperative to foreground participants' voices and not impose an expert account onto their experiences. Yet, as an academic-activist invested in public geographies, I recognise that as a researcher I am part of PAR processes (Fuller & Askins, 2010), engaged emotionally-and-cognitively. Thus, co-production of knowledge involves the minds-bodies of both participants and researchers and requires thinking with and through embodied senses and feelings (see poignant accounts in Moss & Donovan, 2017).

How can we "handle," conceptually and in praxis, the shifting ebb and flow in which rationality-and-emotions are caught up in society and research? Wetherell, McCreanor, McConville, Barnes, and le Grice (2015) argue for a practice viewpoint to examine the entanglements of discourse and embodied feelings. They discount theories of affect as nonrepresentational intensity and only prefigurative to thought and feeling, suggesting such separation of affect:

> from mediated signification is problematic both as a social theory of affect and, particularly, as method [. . . leading] researchers to treat affect as a kind of cultural uncanny: mysterious, a force directly hitting the body, bypassing discourse, sense making and cognition.
>
> (Wetherell et al., 2015, p. 57)

Rather, they argue that any embodied, felt registration of events is simultaneous with meaning making, that emotions are *both* embodied *and* circulated through political and public discourses and practices, co-productive of thought and critical reflection. So, feelings of upset, happiness, disappointment, optimism, and so on sit within sociocultural practices, constructed through dominant norms, with specific limits and articulations. Consequently, feeling-thinking through research processes is likewise entangled in wider structures and practices.

Thus, a specific challenge in PAR is to consider how communities and participants are situated by hegemonic emotional-rational frames that position women/people of colour/ otherwise marginalised groups as responding "too emotionally," "irrationally," to being oppressed. To understand how they themselves might reproduce such frames, *and* how they can challenge and shift these norms. I was challenged as a researcher that "I think too much." Aware of my academic position within hegemonic structures, especially academia as an institutionally powerful whiteness (Muñoz, 2006), I need to progress by:

Situating Feelings

It is both my *desire* and *intellectual training* that require this chapter be placed in empirical context. Feminist and queer theory teaches me that claims to knowledge are being produced *from somewhere*, and *through certain emotions*, which I wish to render transparent here.

When invited to contribute to this book, I was finishing up a three-and-a-half-year project in Glasgow (2015–18). My first instinct was to discuss co-writing with some participants who I thought-sensed were most likely to have an interest, and capacity to do so. Unfortunately, and not for the first time, I discovered that participants had neither, and so I tackled the task alone.[5] However, when rereading field journals across 18 years and multiple projects, I found that memories, experiences, surprises, *and emotions* (mine and participants') came tumbling off the pages. My reflections are intersubjective with a diversity of voices.

Co-writing is a critical part of co-production process, filled with tensions and potential (Nagar, 2014). Yet in much PAR, co-authorship is not possible for various reasons, as discussed further below. So, it is critical to be transparent that analysis, knowledge-making, and dissemination involves researcher reflection. It is inherently, and thus should also be explicitly, about researchers' desire, fear, frustration, anticipation, and so on (Bondi, 2014). Further, these emotions are formed in relation to those we research with.

The projects I've been involved with occurred in deprived areas of Sheffield, Middlesbrough, and Newcastle in England, and Glasgow in Scotland. The shortest project was nine months (a real challenge to PAR!), the longest was five years. As I write, I smile remembering laughter; I frown recalling anger and conflict. Such emotions draw out-and-on the relational geographies of doing research and demonstrate how it is woven and layered through other spaces and times in embodied ways. For example, when explaining the sole authorship of this chapter I still feel the burning sense of disappointment I experienced as a PhD student when I approached key participants to co-write elements of my thesis: "Why would I want to do that? It's your PhD, not mine" was the gist of the responses.

I felt-thought PAR had failed because participants didn't have ownership of the written work. One co-researcher reassured me that they took ownership of the aspects of the project that they wanted to, which were of utility to them. These were mostly action-oriented: funded/organised trips/events, shifts in organisational policy, and working at the local level, and so on. I've since learned to listen more closely to participants' thoughts-and-desires regarding what they want to get from PAR, and have capacity to undertake, and not project my own agenda. I've become increasingly wary of how neo-liberalising universities and funders place pressures on researchers to push participants for evidence of "impact" (see Pain, Kesby, & Askins, 2011).

This need to respect and work with participants' capacity also goes for analysis. It is difficult to build in time, space, and resources (not least payment for participants' efforts) for co-analysis. In a recent Glasgow project, only five of over a hundred participants were involved in analysis, and all were women. McIlwaine and Datta (2004) highlight how wider structural gender norms can reimprint in PAR, especially where a PAR project involves women/more marginalised individuals taking up additional roles as co-researchers. These participants are already doing the bulk of home- and care work, yet are positioned as having time for co-research.

Situating feelings then demand consideration of positionality and power, as embodied and emotional. That is:

Power-ful Feelings

When PAR projects seek to challenge racist, sexist, heteronormative, and other exclusionary worldviews, "maintaining respect for plurality is vital [but] respect does not imply acquiescence" (Wynne Jones et al., 2015, p. 219). Rather, there are deeply contested and difficult power relations caught up in building trust, reciprocity, and respect. How participation is framed, who is involved or left out, involves unpacking how in/exclusionary processes occur and power circulates. Key here is that power operates through emotions-and-affect; structural violence and exclusion *is deeply felt*. Hume (2007) writes that the very emotion of violence is central to understanding how violences work, thus researching issues involving any kind of trauma and oppression must pay attention to the emotionalities in/of social and spatial power relations (see also Lund, 2012; Tamas, 2014). Likewise, understanding how participation is framed and evolves requires attending to the emotional dimensions of in/exclusion and power inequalities.

Further, in trying to address such inequalities, recognising agency is pivotal. It is important to decentralize oppression as the only way to study "otherness" in PAR, and work with participants as they develop their capacities and critical consciousness (see Freire [2007] on PAR and critical pedagogy). Payson (2018) argues the need to "track feelings" to uncover how moments of engagement and communality arise and dissolve in struggles for justice and in research with diverse, marginalised communities. Like Payson, I'm passionately invested in how "feeling together" holds the potential to open up radical spaces of hope and transformation, through research and beyond. And I'm inspired by Mirza (2015), who writes about spaces of radical black agency as forms of collectivity. Such spaces enable richly political activism, precisely through the vital role emotions play in moving beyond prescriptive power structures and discourses – what Mirza (2015) calls "a "quiet riot" that is overlooked in masculine theories of social change" (p. 5). Just as with oppression, then, emotions are central in enacting agency, in opening up ways of "living otherwise" (hooks, 2000) and in progressive forms of solidarity and "emotional citizenry" (Askins, 2016).

To understand the emotional geographies of power in PAR processes, and the communities and participants we research with, it is vital to appreciate the ways in which emotions themselves are socially and culturally constructed in normative discourses and power relations. Feminist, queer, decolonial, and other critical scholarship calls out how institutional and discursive framings of emotions as "private," "feminine," and "exceptional" from the "usual" business of state/politics/citizenship serve to reinscribe existing power relations in patriarchal capitalist worldviews. Having and showing emotions is othered, and logic is normalised/exalted.

Even when emotionality is recognised/legitimised, secular, liberal whiteness is a cultural logic that prescribes and regulates "who may feel and express what" (Muñoz, 2006, p. 680). Feeling hurt, fury, and fear may be accepted or possible for certain people and not others. The dominance of Western perspectives on what is the "right" emotional response in any situation often dismisses black and Indigenous peoples' emotional beings/lives/epistemologies (Ngai, 2007). What to do with "ugly" or "bad" feelings (Ngai, 2007), such as bitterness, resentment, and hatred, is a serious and difficult issue. PAR must contend with who gets to feel what, and whose feelings matter?

This point connects with debates around everyday politics and embodied practices of belonging and encounter, in which desires, fears, affect-and-emotions are paramount (Lobo, 2016; Wise & Velayutham, 2017). Action research that attempts to challenge oppression and work towards emancipatory change is inherently embedded in community and requires emotional intelligence. Participants come together as a research community, working with already-formed communities, and build solidarities and links across groups to develop (new) communities. In all of this, community is deeply contested through relations of power, drawing boundaries as well as enabling belonging, crucially through emotional geographies of encounter.

And as researchers pay increasing attention to more-than-human geographies (see Bawaka Country et al., Chapter 3), issues of emotion and asymmetry of encounter must be extended to social-nature relations, and the contingencies and limits of participation for all involved. Franks (2015) outlines a notion of "passionate modesty" as "a shared reckoning of limitations that can enrich the potential" of PAR (p. 237), by foregrounding the complex, contested relations of solidarity and resistance across both animate and inanimate participants (see also Barker & Pickerill, 2019). In recognising the emotionalities of participant agency, PAR must unpick human-centred discourses that privilege people over planet, as well as sexist, racist, xenophobic, ageist, heteronormative, and other such structures of violence and oppression.

Researching and working with power-ful emotions then requires:

Feeling Methodologies

Over the years, colleagues, students, and participants in PAR have often asked, "What methods will we use?" My answer is that a participatory approach precisely leaves that question open, to be negotiated through evolving process with participants, dependent on the research focus and context. There are no technical blueprints for PAR. What I want to emphasise here is that emotions are central in methodological praxis, emotions directly affect knowledge co-production, and thus critical research methods should attempt to work *with and through* emotions.

In a project on recycling and environmental initiatives in Newcastle (2011–2012), we held two workshops to decide project methodology, through which participants (all from black and ethnic minority communities) strongly favoured a questionnaire survey, consisting of mostly quantitative questions (see also Stoudt, Chapter 5). I felt despondent as this decision cohered; I struggle with statistics, epistemologically and in practice! Participants, though, were being given quite a strong steer by the research partner organisation staff, who had gained funding for the project from the city council. Statistical evidence was not a condition of funding, but encouraged and expected by the council, and staff voiced their concerns about delivering a "successful project" repeatedly, fearful that not doing so jeopardised potential future funding in times of austerity cutbacks.

Other emotions were circulating too. In the workshops, participants were anxious about their ability to conduct interviews, and to recruit interview respondents in their communities:

> Bob was adamant in today's session that local residents would reject requests for interview, that they wouldn't consider recycling as their concern, wouldn't have time or see how an interview would help. Jocinta agreed strongly, arguing that if they knock at doors with a questionnaire instead, they'll be seen as having some authority, and people will more likely answer questions this way. Plus questionnaires are "much quicker so you can talk to more people." There was loud agreement. I then asked, yes but you all are interested in environmental issues, living in the same areas and with similar backgrounds as your neighbours, why would they not be interested? Ali replied "well I don't know how to interview, but I can tick boxes on a questionnaire." Again, vociferous agreement and nodding. I think participants are worried both about not getting responses, and the unknown of an interview. I'm wondering if there's some fear too, among those who've had awful experiences with interviews, through claiming asylum and refugee status?
>
> (personal fieldnotes, July 2011)

Regarding the role of emotion in fieldwork, Bennett, Cochrane, Mohan, and Neal (2015) outline a need for "embodied listening" that "weaves through, around and beyond what is immediately heard" (p. 7), including the unspoken. Crucially exploring the feelings that mediate listening, they outline this approach as entwined through intersubjective dynamics, sensuous and embodied. What is vital is the *relationality* of feelings and experiences as connected to those of others, and that the imprints of relations are carried into future experiences, and into the production of knowledge. Such embodied listening is about engaging fully with unfolding narratives, and is relevant to a range of qualitative methods, despite not sharing the same experiences while using those methods (Bondi, 2014). Yet PAR involves such close personal work with participants – to develop aims, conduct research, and check in though analysis, writing, and dissemination stages – that many experiences become shared. Lines are blurred, emotions intersubjective, and experiences built together with participants as co-researchers.

Cahill, Quijada Cerecer, and Bradley (2010) emphasise the multiple, complex, and emotional ways in which new subjectivities evolve in PAR. Personal and social transformation is possible through PAR methodologies *as critical pedagogy* (after Freire, 2007). Activities in which participants investigate their everyday lives are key in identifying personal exclusions as shared, and thus social and political, and such learning is emotional-and-cognitive. Similarly, Bignante, Mistry, Berardi, and Tschirhart (2016) reflect on the role of emotions in shifting self-perceptions among community researchers involved in participatory video, changing attitudes, values, and visions (see also Brickell, 2015).

Recent emphasis on creative methodologies across the social sciences resonates here. Using art, drama, video, and diverse "doings" (Kale, Kindon, & Stupples, 2018; Tamas, 2014; and Shaw et al., Chapter 8) holds potential to open out to emotions both as topic of research, and as woven through research. In a participatory art project, Diprose (2015) found that through making art, participants were able to recognise their own and others' complex subjectivities as both complicit with normative, unjust discourses and structures, as well as desirous for emancipatory change (see also Marnell [Chapter 4] and Woodrow [Chapter 6] in this volume). This arts-based PAR facilitated emotional engagements with art and each other, which served to counter hegemonic Western theories centring a unified subject as prefigurative for political change. Rather, their process of *creating* facilitated participants to articulate their ambivalent and fluid subjectivities in embodied and verbal ways (see Figure 7.1).

I've also found that practice-based methodologies enable participants to explore difficult experiences through thinking-while-doing and emotionally-driven-cognition:

> New connections were forged today between Peter (white, Scottish, 30s, long term resident), Sarah (white, Scottish, 30s, recent resident), Mohammed (black, Sudanese, 50s, refugee) and Samira (black, Eritrean, 40s, refugee). While making bread, Peter discussed being "sanctioned," his state benefit payments were withdrawn when he missed an appointment at the job centre (he was in hospital). Everyone agreed how

Figure 7.1 Recognising complex subjectivities. Courtesy of Magda Kaminsha, 2017

> unfair this was, how uncaring and inhumane the "system" is. Then Samira mentioned difficulties in getting state housing benefit once she was granted refugee status, being forced to live in a homeless hostel for a month. She said she felt so ashamed to be asking for money, she'd rather pay her own rent but [UK regulations meant she] couldn't work as an asylum seeker then found it hard to get a job once refugee status was granted. Mohammed told a similar story. Peter agreed, saying all he wanted was a job but being in recovery from substance misuse made it impossible, and "they make you feel so small like you're nothing you don't deserve a chance."
>
> There was quiet for a while, as we kneaded the dough. Then Mohammed started to dance, saying if you dance as you knead more energy goes into the loaves . . . we joined in and everyone was giggling. Sarah then said how the project [run by research partner organisation] gives her a sense of pride, said she felt ashamed too at not being able to find work after being at home bringing up her kids: "I've no education and they look at you at the job centre like you're stupid, but I've got skills." Peter said Sarah and Mohammed were excellent bread makers, Sarah then complimented Samira on her "brilliant biscuits."
>
> (personal fieldnotes, November 2017)

In this fieldwork moment, participants articulated and shared feelings of shame that arose through structural exclusions, while also supporting each other by empathising, by complimenting, and helping each other physically (to make the bread). "Talking" research methods, such as interviews or focus groups, may not have elicited such empirical data. Further, trust had been developed through the PAR process, such that people felt comfortable discussing intimate experiences. The informality of method was important too, I'm sure. Somehow, the mundane act of making bread facilitated emotional connections to come to verbal cognition (see Figure 7.2).

PAR is about *learning with* participants and *through relations* to co-produce knowledges and actions. Methodological approach should then consider what types of relations are being produced for whom and why, by paying attention to the various emotional dimensions of generating data and knowledge. I believe that emotions can be better incorporated and listened to, so that PAR is more fully a collective imagining and enacting of new ways of living, by taking seriously and hopefully the:

Emotional Geographies of PAR

In this chapter, I've tried to convey that situated and power-ful emotions circulate through the relational geographies of PAR, co-producing knowledges and holding the potential to create new politics of affirmation and anticipation, as felt-and-thought (see Boggs, 2012 regarding *evolution* beyond revolution). Through reflection across fieldwork projects, I've tried to show that emotions are inherently and explicitly involved in all aspects of participation, action, research, and representation.

I want to turn now to how feelings engendered through PAR affect other aspects and spaces of my life, whether formally captured as empirical evidence or not. Years of accumulated emotional encounters feed into my ongoing activities and approaches in research, teaching, and personal geographies, and all kinds of transformations beyond the scope of any specific research project.

In particular, I want to stress the role of emotions in shifting academic orientations, with regards research-led teaching, and the university as site of participatory pedagogy.

Figure 7.2 Practice-based methods. Courtesy of High Rise Bakers, 2018

Mountz, Moore, and Brown (2008) discuss the multiple relational geographies of PAR and/as community-based "service learning," through which students learn in the classroom and through action research placements. Adopting a participatory frame requires students, organisational, and academic staff to listen to and work carefully with one another, alongside people involved with whatever project evolves. Simultaneous intellectual-and-emotional student learning occurs through process, iterative ethics, and *developing relationships* (as with PAR itself, see Blazek & Askins, 2020). The value of PAR-placement modules cannot be underestimated: time and again undergraduate students I've worked with in such ways describe having learned so much from working together with people and in communities, "more than being in the classroom" (various, personal communications).

PAR can also foster relationships that bring co-researchers into the classroom. After several months working together on a project in Newcastle (2007–2010), Hassan, a Rwandan man, suggested that he talk with students on the Geopolitics of Ethnicity module I taught. For seven years he gave an annual session, in which he spoke about his personal experiences as a refugee, moving across Africa and Europe, claiming asylum in the UK, and settling in Newcastle.[6] Trained as a lawyer, and employed as a project officer supporting asylum seekers and refugees, he was also able to give details that I could not about the wider issues, institutional politics, and public discourses that impacted his life. Students always responded emotively-and-thoughtfully to Hassan and his talk, learning beyond any lecture/workshop/session I could deliver regarding these issues and geographies.

Let me end by returning to the opening moments and title of this chapter. It isn't that as researchers/students/participants we can ever think too much. Rather, by also wearing our hearts on our sleeves we can seek to better interconnect logic-emotion and mind-body; and in so doing, crucially enact research and university spaces in more emancipatory ways. Recognising the colonial logics circulating in education itself, in an "era of re-colonisation" (Noxolo, 2015), there's a critical need to *work within-and-without academia*, through recognising emotions as already present, and consciously working with emotions. It is my experience that thinking-emotionally/emotional-intelligence can enable connections across radical spaces of hope and transcendence, through research and beyond. Such connections and moments are fragile, and I'm wary of overplaying their potential, yet PAR can render new ways of being-becoming-and-feeling together; new political languages and actions that surpass patriarchal capitalist coloniality.

I've learned so much through PAR over the years. It is unpredictable, messy, unruly, contested, exhausting, rewarding, fluid, and surprising. It's ***always emotional: joyful, exasperating, full of hope, anger, regrets, love, and care*** . . . Being heedful of the emotional geographies of PAR enriches research and enhances intellectual rigor. Holding onto emotion is another dimension of critical reflexivity and situated knowledge-making, that enacts more robust science than ignoring it. Recognising emotion likewise enriches our selves, as thinking-feeling-becoming with others.

Acknowledgements

I am deeply indebted to so many people who, over the last 18 years, put their time, energies, thought, and care into research and action with me. I'm humbled. It's been my honour, struggle, and pleasure to share innumerable journeys and experiences. I'm hopeful that this narrative in some small way conveys the countless lessons I've been taught.

I also offer heartfelt thanks to Sara, Mike, and Rachel, who have inspired me from my PhD to the present day. So too have the passionate pygyrg-ers (members of the Participatory Geographies Research Group [https://pygyrg.org/]) and other critical academics it has been my fortune to encounter, debate, and socialise with, and the students I've been privileged to learn alongside. The late Duncan Fuller continues to inspire my writing.

Notes

1 Identifications were recorded at the time, as pertinent to the focus of each project; pseudonyms are used throughout this chapter.
2 To "wear your heart on your sleeve" means to "to allow your feelings to be obvious to everyone around you" ("Wear your heart on your sleeve," n.d.).
3 I hyphenate volunteer-researcher to indicate a fluid and complex role in which I was both a researcher and acted as a volunteer; not either/or, but aspects of both that ebbed and flowed across space and time. To be clear, I was a salaried academic at all times.
4 I hyphenate *know-feel* (also *mind-body, emotions-and-affect, emotional-rational*, etc.) to try to convey connectedness over separateness. While a slash (/) can convey *either/or*, I mean *both-and*, as central to the intention of the chapter. This hyphenating remains problematic, on purpose: it retains some separation, which points to the difficulties of properly conceiving beyond binaries in Western, modern academic thought.

5 I have often deployed "*et al.*" in authorship to indicate inherent co-authorship, mindful of the multiplicity of voices, I have a responsibility to, yet can never "represent."
6 For which Hassan was paid an hourly lecturer rate by the university. There are critical issues around resourcing participants for time and effort.

References

Askins, K. (2016). Emotional citizenry: Everyday geographies of befriending, belonging and intercultural encounter. *Transactions of the Institute of British Geographers*, 41(4): 515–527. https://doi.org/10.1111/tran.12135.

Barker, A. J., & Pickerill, J. (2020). Doings with the land and sea: Decolonising geographies, Indigeneity, and enacting place-agency. *Progress in Human Geography*, 44(4): 640–662. https://doi.org/10.1177%2F0309132519839863.

Bennett, K., Cochrane, A., Mohan, G., & Neal, S. (2015). Listening. *Emotion, Space and Society*, 17: 7–14. https://doi.org/10.1016/j.emospa.2015.10.002.

Bignante, E., Mistry, J., Berardi, A., & Tschirhart, C. (2016). Feeling and acting 'different' emotions and shifting self-perceptions whilst facilitating a participatory video process. *Emotions, Space and Society*, 21: 5–12. https://doi.org/10.1016/j.emospa.2016.09.004.

Blazek, M., & Askins, K. (2020). For a relationship perspective on geographical ethics. *Area*, 52(3): 464–471. https://doi.org/10.1111/area.12561.

Boggs, G. L. (2012). *The Next American Revolution: Sustainable Activism for the 21st Century.* Berkeley: University of California Press.

Bondi, L. (2014). Understanding feelings: Engaging with unconscious communication and embodied knowledge. *Emotion, Space and Society*, 10: 44–54. https://doi.org/10.1016/j.emospa.2013.03.009.

Brickell, K. (2015). Participatory video drama research in transitional Vietnam: Post-production narratives on marriage, parenting and social evils. *Gender, Place & Culture*, 22(4): 510–525. https://doi.org/10.1080/0966369X.2014.885889.

Cahill, C., Quijada Cerecer, D. A., & Bradley, M. (2010). "Dreaming of. . . ": Reflections on participatory action research as a feminist praxis of critical hope. *Affilia: Journal of Women and Social Work*, 25(4): 406–416. https://doi.org/10.1177%2F0886109910384576.

Davidson, J., Bondi, L., & Smith, M. (2014). An emotional contradiction. *Emotion, Space and Society*, 10: 1–3. https://doi.org/10.1016/j.emospa.2013.12.014.

Diprose, G. (2015). Negotiating contradiction: Work, redundancy and participatory art. *Area*, 47 (3): 246–253. https://doi.org/10.1111/area.12177.

Franks, A. (2015). Kinder cuts and passionate modesties: The complex ecology of the invitation in participatory research. *Area*, 47(3): 237–245. https://doi.org/10.1111/area.12171.

Freire, P. (2007). *Pedagogy of the Oppressed.* New York: Continuum.

Fuller, D., & Askins, K. (2010). Public geographies II: Being organic. *Progress in Human Geography*, 34(5), 654–667. https://doi.org/10.1177%2F0309132509356612.

hooks, b. (2000). *All About Love: New Visions.* New York: HarperCollins.

Hume, M. (2007). Unpicking the threads: Emotion as central to the theory and practice of researching violence. *Women's Studies International Forum*, 30(2), 147–157. https://doi.org/10.1016/j.wsif.2007.01.002.

Kale, A., Kindon, S., & Stupples, P. (2018). 'I am a New Zealand citizen now—This is my home': Refugee citizenship and belonging in a post-colonizing country. *Journal of Refugee Studies*, 33 (3): 577–598. https://doi.org/10.1093/jrs/fey060.

Lobo, M. (2016). Geopower in public spaces of Darwin, Australia: Exploring forces that unsettle phenotypical racism. *Ethnic and Racial Studies*, 39(1): 68–86. https://doi.org/10.1080/01419870.2016.1096407.

Lund, R. (2012). Researching crisis—Recognizing the unsettling experience of emotions. *Emotion, Space and Society*, 5: 94–102. https://doi.org/10.1016/j.emospa.2010.09.003.

McIlwaine, C., & Datta, K. (2004). Endangered youth? Youth, gender and sexualities in urban Botswana. *Gender, Place and Culture*, 11(4): 483–512. https://doi.org/10.1080/0966369042000307979.

Mirza, H. (2015). "Harvesting our collective intelligence": Black British feminism in post-race times. *Women's Studies International Forum*, 51: 1–9. https://doi.org/10.1016/j.wsif.2015.03.006.

Moss, P., & Donovan, C. (Eds.) (2017). *Writing Intimacy into Feminist Geography*. London: Routledge.

Motta, S. C., & Seppala, T. (2016). Feminized resistances. *Journal of Resistance Studies*, 2(2): 5–35. https://resistance-journal.org/.

Mountz, A., Moore, E. B., & Brown, L. (2008). Participatory action research as pedagogy: Boundaries in Syracuse. *ACME: International E-Journal for Critical Geographies*, 7(2): 214–238. https://acme-journal.org/.

Muñoz, J. E. (2006). Feeling brown, feeling down: Latina affect, the performativity of race, and the depressive position. *Signs: Journal of Women in Culture and Society*, 31(3): 675–688. https://doi.org/10.1086/499080.

Nagar, R. (2014). *Muddying the Waters: Coauthoring Feminisms across Scholarship and Activism*. Champaign, IL: University of Illinois Press.

Ngai, S. (2007). *Ugly Feelings*. Cambridge, MA: Harvard University Press.

Noble, G. (2009). Everyday cosmopolitanism and the labour of intercultural community. In A. Wise & S. Velayutham (Eds.), *Everyday Multiculturalism*. 46–65. Basingstoke: Palgrave Macmillan.

Noxolo, P. (2017). Decolonial theory in a time of the re-colonisation of UK research. *Transactions of the Institute of British Geographers*, 43(2): 342–344. https://doi.org/10.1111/tran.12202.

Pain, R., Kesby, M., & Askins, K. (2011). Geographies of impact: Power, participation and potential. *Area*, 43(2): 183–188. https://doi.org/10.1111/j.1475-4762.2010.00978.x.

Payson, A. (2018). *Feeling Together: Heritage, Conviviality and Politics in a Changing City*. PhD thesis. Cardiff: Cardiff University. https://orca.cardiff.ac.uk/108561/.

Tamas, S. (2014). Scared kitless: Scrapbooking spaces of trauma. *Emotion, Space and Society*, 10: 87–94. https://doi.org/10.1016/j.emospa.2013.08.001.

Wear your heart on your sleeve (n.d.). Collins Dictionary. https://www.collinsdictionary.com/dictionary/english/wear-your-heart-on-your-sleeve.

Wetherell, M., McCreanor, T., McConville, A., Barnes, H. M., & le Grice, J. (2015). Settling space and covering the nation: Some conceptual considerations in analysing affect and discourse. *Emotions, Space and Society*, 16: 56–64. https://doi.org/10.1016/j.emospa.2015.07.005.

Wise, A., & Velayutham, S. (2017). Transnational affect and emotion in migration research. *International Journal of Sociology*, 47(2): 116–130. https://doi.org/10.1080/00207659.2017.1300468.

Wynne-Jones, S., North, P., & Routledge, P. (2015). Practising participatory geographies: Potentials, problems and politics. *Area*, 47(3): 218–221. https://doi.org/10.1111/area.12186.

8 Pathways to Scaling Social Inclusion Innovation Through Participatory Action Research

Jackie Shaw, Sowmyaa Bharadwaj, Anusha Chandrasekharan, and Dheeraj

The Participate Initiative developed scaling methodologies during Sustainable Development Goal (SDG) formation and implementation. Scaling strategies included: purposively networking 18 Participatory Action Research (PAR) partners; nurturing inclusive relations, iterative learning, and community emergence through PAR; using innovative visual methodologies; and building dialogic spaces with policymakers at national and global levels. We exemplify these strategies through our collaborations using participatory video alongside other methodologies in three Indian projects. As anticipated, there were inherent tensions, with scaler analysis highlighting the complication and compounding of constraining power dynamics as collective action and deliberative spaces diversify. However, counterintuitively, we found using visual methodologies for PAR contributes by making these tensions more *visible*. We present our key learnings about the necessary negotiations to inform inclusiveness in future scaling projects.

Introduction

Global inequality is rising along with the closure of civic space. Despite large development investments, insecurity and exclusion are increasing for the poorest people (Burns, Howard, Lopez-Franco, Sharokh, & Wheeler, 2013). Right-wing populism and nationalist politics are gaining prominence in many places North and South. This leads governments to turn their backs on international cooperation and abrogate responsibility for marginalised people, including welfare recipients, street-dwellers, refugees, and sexual minorities (Wheeler, Shaw, & Howard, 2020). Meanwhile, the climate emergency starkly illustrates the co-constituting link between global governance failings and local environmental impacts. Although the current situation is depressing, and there is an urgent need to realistically assess improvement efforts and the magnitude of challenges, it is also vital to nurture a politics of hope.

An alternative to the anti-internationalist popularist trend is the global "leave-no-one behind" agenda (Kabeer, 2016). However, this cannot be achieved without learning with people in situ about the barriers to inclusion (Howard, López Franco, & Shaw, 2018; Shaw, Howard, & López Franco, 2020). "Social inclusion innovation" is defined as the development of new knowledges and practices to improve social, economic, health, and political conditions for disadvantaged and excluded groups (George, McGahan, & Jaideep, 2012). "Scaling" for social inclusion innovation is concerned with optimising the social impact, or the quality, extent, and spread of beneficial effects in marginalised people's lives (McLean & Gargani, 2019). These processes raise key questions: What is being

DOI: 10.4324/9780429400346-8

scaled and why? and How can different scaling actions with different foci and directions operate effectively?

Participatory Action Research (PAR) offers a means to cultivate *social inclusion innovation* because knowledge on practical solutions can be generated by involving participants in trial-and-error learning as they tackle their own issues. Indeed, there are many examples of successful local changes through PAR interventions (Howard et al., 2018; Kindon, Pain, & Kesby, 2007, Chapter 1). In theory, progressive cycles of longer-term PAR also provide the potential to scale their change processes, and this has been achieved to some extent in some circumstances (Burns & Worsley, 2015; Shahrokh & Wheeler, 2014; Shaw, 2017, 2021). Nevertheless, scaling of insights and participant-led action is often limited by time, resourcing, and co-option by powerful interests, which all dilute their transformative promise. Thus, scaling PAR up and out to increase social impact can prove difficult in practice.

This chapter explores the possibilities and challenges of using PAR to scale down to generate *social inclusion innovation*, and then to scale out and up the knowledges and inclusive practices developed through innovative networking and communications strategies. Using scale as a conceptual lens to clarify the progressive tensions faced when scaling PAR down, out, and up, we reflect critically on the connections between temporal and spatial scales, and the interplay between different scaling activities.

To do this, we draw on our experiences during the global research programme Participate,[1] which between 2012 and 2018 developed and applied a "scaling pathways" methodology. Participate involved a network of 18 partner organisations using long-term PAR processes with highly marginalised groups in 30 countries. Steered by a team from the global South in collaboration with the Institute of Development Studies (IDS), Participate used a "shared learning" and "collective leadership" model. This model pooled skills, intentionally connected locally led PAR processes across the multiple contexts, and leveraged the network's learning in order to influence policy-making and programme implementation. Jackie is a research fellow at IDS and led the Participate visual methodologies programme. Sowmyaa, Anusha and Dheeraj work for *Praxis* in India, a key Participate steering partner.[2] Full details of Participate's activities are beyond the scope of this chapter. Here we draw specifically on our direct collaboration in three Indian contexts, whilst acknowledging the wider network collaborations which contributed to our insights.

The next two sections frame the empirical discussions by introducing scale and scaling terminology. They problematise PAR's use for generating *social inclusion innovation* and outline recent thinking on how to scale out and up more effectively. Section 3 introduces the main scaling strategies developed during the Participate programme. Reflecting on our collaborations, section 4 illustrates how these strategies played out during three different projects in India, focusing on participatory video's contribution, alongside other methodologies, within the wider *Praxis* PAR processes. Finally, we summarise key learning about how to navigate the inherent challenges on the pathways to scale from the social margins.

Conceptualising Scale and Scaling Terminology

Scale is socially produced and predominately epistemological (Jones, 1998). Different analysis scales, from macro-level to micro-level, or from global to local, are different "ways of knowing," which generate different insights on social domains, phenomena, or processes. Therefore, approaching the relationship between these categories as ontologically existent and vertically fixed, with hierarchical scale units, functions problematically. Such an approach positions the global (universal) as more important than the local (particular). It makes scaler units appear natural and neutral and makes the adaptive and constantly evolving world seem permanent and unchangeable. Vertically nested scaler imaginaries prove inadequate for understanding a complex, interconnected world (Brenner, 2001), and make it harder to envision routes towards more equal societies.

Scale concepts also have actual ontological effects, in that they impact what can happen in reality. Global, national, and local scales may be social constructions (Marston, Jones, & Woodward, 2005), but they have real boundaries that make it difficult to tackle phenomena that function across them (e.g., the environment). Thus, the contemporary perspective is that scale can neither be approached as entirely ontological (actual and permanent) nor wholly epistemological (created discursively). In this sense, scale is always political, as it delineates the virtual imaginaries of power, and the spaces for and processes of resistance (Smith, 1992). A global perspective that does not prioritise any spatial orientation is needed (Sassen, 2008), to understand the nuances of how the global manifests locally, how local forces (re) produce the global (Darian-Smith & McCarty, 2017), and how new social possibilities can emerge in context. Therefore, scaler theory has been reorientated horizontally as complex networked relations that flow across and between binary scale divisions (Amin, 2002). This gives scale dynamic and relational dimensions and adds temporal as well as spatial scales into the mix.

Finally, there are subtle differences in scaling terminology in different academic disciplines and practice contexts. Broadly, scaling for social change in community and development settings refers to maximising beneficial impacts, by shifting the usual scaler focus or practice in a particular direction (e.g., down, up, in, out, deep, or long) (Cohen & McCarthy, 2015; McLean & Gargani, 2019). However, PAR processes are often conflated with scaling down alone.

Scaling down here refers to situating activities such as knowledge production, governance decision-making, or improvement action locally, which may be defined by place (e.g., a neighbourhood, village, or district), or by a particular subgroup in a locality. The assumption is that scaling down will decentralise influence and empower local actors. However, operationalising scaling down through participatory approaches can also disempower, as well as unjustly shift responsibility for solving systemic issues to local people (Cohen & McCarthy, 2015). Other scaling terms have arisen to frame more empowering approaches (McLean & Gargani, 2019), such as *scaling long*, or *temporal scaling* through longer-term engagement; *scaling deep*, through the type and quality of relations and processes needed for meaningfully inclusive participation; and *scaling in* through building organisational capacities and improved practice.

Scaling out refers to extending the reach of participatory projects by repeating an intervention with similar groups at additional sites over a larger area. *Scaling out* also refers to communicating thematic or methodological learning more widely, which diffuses knowledge to benefit more people or engages other groups in action on an issue. This kind of scaling out is called *horizontal scaling*, if it involves groups that are similarly positioned locally due to socioeconomic or identity factors, either through the original project drawing in more participants across a wider area or through project repetition in other localities, without being joined to one another as a regional or national initiative. *Diagonal scaling* (Fox, 2015) involves connecting participants with more influential local stakeholders, such as service providers, business actors, or nongovernmental organisations (NGOs) operating nearby. *Scaling up*, or *vertical scaling*, involves leveraging local knowledge or collective action practices to influence national or international policy and programmes through advocacy, networking, negotiation, and partnership. *Scale jumping* disrupts the normative scale at which an activity usually operates, by applying it at a different scale. Scaling up can involve scale jumping when participatory approaches usually used at ground level are used at a policy level. As with scaling down, these scaling activities can both improve and worsen circumstances for the most excluded groups (Cohen & McCarthy, 2015).

Despite extensive debate in disciplines such as human geography, anthropology, and politics over the last few decades, what scale means remains contested (Escobar, 2007; Leitner & Miller, 2007; Marston et al., 2005). Scaling literature also reveals a need for clarity about *what* is being scaled, which might be knowledge (e.g., thematic insight or learning about change enablers and barriers), approaches or mechanisms (e.g., participatory practices, research methodologies, communication processes), or policy-oriented outcomes. This breadth of possibilities has resulted in a plethora of new scaling action terms, including *scaling methodologies, scaling agency, scaling influence*, or *scaling sustainability*.

A concept's usefulness lies in what it contributes to empirical insight and guiding action, and some argue that scale has lost utility (Marston et al., 2005). Certainly, academic debates have clarified the need to think carefully when applying scale terms and scaling processes. In this spirit, we apply spatial and temporal scaling lenses to explore how they illuminate the use of PAR as a pathway to scaling social inclusion innovation. The next section focuses on the multiple scaler processes involved in PAR, to think more critically about scaler politics and dynamics.

Using PAR for Scaling *Social Inclusion Innovation* – Key Issues and Reorientation

PAR is a long-established metamethodology for tackling stubborn social problems (Kemmis & McTaggart, 2005; Reason & Bradbury, 2008). It can democratise knowledge production processes by scaling down to involve people facing an issue in interpreting their own situations (Chambers, 1997). Prioritising these agendas can foster more equitable and trusting research relationships, which are necessary for deeper understanding (Gaventa & Cornwall, 2008; Park, Brydon-Miller, Hall, & Jackson, 1993). Crucially, PAR can be an engine of innovation because it drives iterative cycles of improvement action, evaluation, and adaptation (Burns & Worsley, 2015; Kindon, Pain, & Kesby, Chapter 1). However, there are considerable gaps between PAR ideals and realities, which scaler concepts help clarify.

The term *participation* is polyvalent (Kesby, 2007). It is sometimes deployed to inspire social justice action, but can be co-opted (e.g., by the state, intergovernmental organisations, NGOs, and universities) to serve mainstream neoliberal agendas such as cost-effectiveness (Mansuri & Rao, 2004). State-led scaling down of governance decision making through participation can be operationalised through shallow or short-term consultation processes in governance-led spaces. However, less tokenistic participatory processes can still pass responsibility for services to citizens, and to NGO or private-sector actors, which often leads to victim blaming and additional burdens for those worst affected. In many cases, power shifting rhetoric masks neoliberal agendas (Cooke & Kothari, 2001), while the real power remains with the state (Cohen & McCarthy, 2015).

Similarly, PAR's transformational possibilities have often been diluted through institutionalisation in academic, local governance, and NGO contexts. The rapid proliferation of Participatory Rural Appraisal (PRA) as quick-fix intervention (Chambers, 2005) illustrated how PAR can lose quality when processes that should be contextualised and emergent are instrumentalised and scaled out. PAR is often most effective and ethical when it is small scale and spatially localised, while scaled long through extended time periods. This allows for progression beyond single-loop problem solving, which merely fine tunes current practices, and enables double-loop learning to critically question the assumptions behind accepted understanding (Maurer & Githens, 2009). However, too frequently PAR is operationalised at time scales too short for iterative learning processes, which limits opportunities to scale deep and evolve towards transformative insight and sustainable action.

Scaling down and out through locally led PAR processes (whether instigated by local state actors or community-based NGOs) also often fails to be inclusive (Cohen & McCarthy, 2015). Marginalisation of particular people in situ is often a historical process based in intransigent social norms. Approaching communities as if they are homogenous masks relative disadvantage and excludes those with low status (Dinham, 2005). Indeed, the ways structural power is sustained through microlevel interactions (Foucault, 1980) within a locality is one of the primary challenges for inclusive PAR. A scale unit like "community" can seem to legitimise a PAR process, while community organisations and leaders can reinforce their entrenched power dynamics (Herbert, 2005). To build inclusive PAR processes, it is first necessary to recognise that marginalised "communities" often do not preexist (Shaw et al., 2020; Watt, 1991), and to invest time in engaging and building the capacities of people who would not usually take part (Howard et al., 2018). Second, nurturing shared purpose within groups while recognising intersectional differences between participants requires sensitive and power-aware facilitation (Ledwith & Springett, 2010), including active input from facilitators to generate inclusive relational dynamics (Shaw et al., 2020; Wheeler et al., 2020). Third, as group processes unfold and scale out to engage wider stakeholders, care is needed to maintain inclusive communication, while building collective action. Without these, PAR easily strengthens existing local hierarchies and perpetuates marginalisation. Addressing these issues is part of scaling deep. Finally, to sustain inclusiveness, organisational input is needed (scaling in) to contextualise and embed inclusive participatory methodologies and the tacit relational aspects behind such methodologies' successful application. This input affects resourcing (financial scale) and time allocation (temporal scale).

With PAR's methodological scale assumed to be local, PAR is commonly qualitative, and the evidence generated is therefore often critiqued as anecdotal without wider applicability. However, methods and scale are not necessarily synonymous. Democratised quantitative methods can involve local people analysing national data or increasing methodological scale through extensive data collection on local issues, such as with participatory statistics (Stein & Jaspersen, 2021; Stoudt, Chapter 5). Importantly, qualitative PAR need not be locally focussed. Building deep understanding from their lived experiences, participants can also explore how regional, national, or global issues play out locally. Such analysis can generate high-quality knowledge about underlying root causes and structural power dynamics, which are crucial to generating transformative solutions and avoiding the unintended consequences of top-down interventions (Burns et al., 2013; Shaw, 2015). These possibilities strengthen the argument that intractable problems in the most inequitable contexts are best tackled by scaling down through small-scale PAR in local sites – that is, if processes are extended temporally (scaled long) for critical learning and inclusion purposes (scaling deep and in), in order to generate *social inclusion innovation*. However, participatory researchers should not be fixated on the local, and must find ways to engage with wider-scale structural inequalities. Scaling out locally generated insights to identify wider resonance across context, and scaling up to influence policy, can assist PAR to maximise the contribution to tackling national or global injustice.

That said, scaling up social inclusion innovation cannot be approached through commercial methods (McLean & Gargani, 2019), and requires alternative approaches (Burns & Worsley, 2015). Neither can best practice be simply scaled up or out through expanding programmes (Narayanan, Bharadwaj, & Chandrasekharan, 2015), because scale ontologies are materially grounded in specific places, and the contextual differences mean what works in one setting may not in another. Broad scale participatory discourse and processes can neglect the way system dynamics are changed by scale, as well as the care of local actions, and these neglected aspects can cause the rollout of pilots to fail (Burns & Worsley, 2015), as well as raise questions of justification and optimal scale (McLean & Gargani, 2019). Tackling microlevel dynamics through participatory group processes is different from tackling meso- or macro-level power imbalance towards wider transformation. Scaling activities necessitates not only revealing but also negotiating intransigent power dynamics, which is more complex at scale.

In consequence, thinking has shifted on what scaling for social impact means, and how it can be practically achieved. Social researchers can be viewed as innovators who evolve novel solutions to tenacious social problems (McLean & Gargani, 2019). Burns and Worsley (2015) apply complexity theory and associated knowledge on how change happens to suggest that bigger is not necessarily better when scaling social affect, and that multiple small interventions in the right place are more likely to shift system dynamics. They also argue that fostering pathways to scale through PAR is more akin to social movement building, where multiple activities are seeded and take hold due to local ownership. From these perspectives, scaling involves first nurturing improvements through small, targeted interventions and dynamic adaptation, to evolve and embed practices, which can then be supported to grow and spread. Burns and Worsley (2015) propose that accelerating contextualised uptake can be achieved by channelling synergistic energy through networks. Scaling social inclusion innovation out and up is thus a complex and nonlinear journey travelled by collaborating social innovators, which McLean and Gargani (2019) conceptualise as navigating the *pathways towards scale*. The iterative action-reflection cycles of longer-term PAR can be a way to learn how to navigate scale from the bottom up. However, these PAR processes must progress beyond the group or local level, to address the common dilemmas and recurrent challenges that prevent scaling out and up. These processes must also maintain the attention and time investment needed to build inclusive dynamics, deep learning, and quality processes.

Overall, there is a compelling case for a broader PAR orientation, through scaling up and out the grounded insights generated, as well as the processes that build sustainable inclusion and drive collective action. This can be done in many ways, including through exploring national and international issues locally, as well as by opening dialogic space across context and levels and by bringing local perspectives into international policy space. In the next section, we introduce the Participate programme context where some key strategies for scaling PAR more effectively and ethically were tested.

The Participate Initiative: Scaling Socially Inclusive Knowledge and Practices through Networked PAR Processes

Participate phase 1 (2012-2014) (see López Franco & Shahrokh, 2024) was initiated to bring lived experiences of poverty into the UN deliberations that informed the Sustainable Development Goals (SDGs) (Burns et al., 2013, Narayanan et al., 2015). Later phases of Participate (2015 to 2018) (see Howard & Shaw, 2024) researched how to build sustainable inclusion and participatory accountability during SDG implementation (Howard et al., 2018; Shaw et al., 2020). The first phase was an ambitious attempt to learn collectively from the grounded insights generated by spatially local and temporally long PAR processes across the network, and then scale up by bringing these knowledges into global policy processes. Everyone involved appreciated the challenge, not least because only 15 months remained before the final UN SDG summit. Some considered the SDGs a distraction because exclusionary dynamics are more entrenched at global scale. The IDS team was mindful of the Voices of the Poor programme (Rademacher & Patel, 2002), where decontextualised use of participant quotes by policymakers generated a participatory facade to obscure neoliberal agendas. Notwithstanding this, IDS and the Participate steering partners considered this initiative a seize the moment opportunity to influence global policy (Shahrokh & Wheeler, 2014).

Participate research design was emergent rather than predetermined but followed some core scaling principles as summarised in Table 8.1.

The first strategy in scaling influence was to purposefully convene a network of partners who were already conducting self-financed long-term PAR with highly marginalised groups (Burns & Worsley, 2015). This meant network learning was based in ongoing participatory relationships (scaling long), to avoid tokenistic participation and extractive shallow research processes. The overall research question asked how change can happen and the barriers to change in context. This was broad enough to enable participants to explore what mattered to them, which was crucial to energising self-directed local action as a precondition for scaling deep and out. The research question also meant the processes had inherent local relevance regardless of what was achieved globally. The question framing also provided an opportunity to explore resonance and difference across a diverse range of inequitable and unaccountable contexts. Thus, IDS had other roles facilitating learning across the multiple inquiry streams in the network, building spaces for dialogue between social levels, and creating communication processes to link the networked PAR learning with global policy deliberations. Finally, the potential of partner's local PAR processes was extended through IDS training input and direct accompaniment in using visual methodologies such as participatory video, digital storytelling, and photovoice.

While seeing opportunities in these strategies, the IDS team and network partners presumed that fostering pathways to scale would not be straightforward, linear, or ideal (Howard et al., 2018; Shaw, 2015), because collaborations existed in contested contexts and there would be unavoidable tensions in trying to shift communication dynamics across levels. It was anticipated that crucial learning would emerge from surfacing these dilemmas and building knowledge about how to *navigate the pathways to scale* (McLean & Gargani, 2019) given the necessary risks and likely trade-offs. The next section exemplifies these scaling strategies.

Table 8.1 Strategies for scaling *social inclusion innovation* through PAR

Strategies	*Principles/elements of the approach*	*Potential scaling contribution*
Purposeful networking across diverse global contexts	• Builds on long-term relationships network partners have with marginalised groups • Increases broader collective learning opportunities • Global South steering group and distributed leadership	• Scaling long and deep to increase potential of local PAR processes • Creates space, relations, and mechanisms for scaling up • Fosters preconditions for larger-scale social impact from PAR
Nurturing *social inclusion innovation*	Using participatory video with other methods to: • Reach and motivate participants • Create enabling, interactional spaces and inclusive dynamics • Focus on local priorities – broad research question • Build local power – agency and collective identities	• Scaling down to generate grounded knowledge on social issues and inclusive participatory practices • Scaling long and deep for meaningfully inclusive PAR processes to address constraining power dynamics • Fostering local ownership as basis for scaling influence and sustainability
Iterative action-learning cycles to deepen insight and improve practice	• Iterative action-learning cycles for deeper criticality on issues and solutions • Use of storytelling to reveal systemic aspects and envision change pathways • Methodological layering (multiple ways of seeing)	• Scaling deep (and long) for critical knowledge of the contextual change drivers and barriers
Fostering community emergence through PAR	• Building collective identities • Group ownership drives exploratory inquiry and collective action • Building solidarity between diverse people to increase power to leverage governance response	• Scaling out to bring diverse people together to find common interest and purpose • Scaling up by finding shared identities at a wider scale
Using innovative visual methodologies	• Training input and direct accompaniment support • Use visual methodologies to drive and mediate PAR • Create visual products to bring participants perspectives to local, national and global space • Visual methods to increase audience empathy and compel action	• Scaling in to embed inclusive practices • Support scaling deep inclusively when process is temporarily long • Means for scaling up research insights on issues and practices, and policy recommendations
Building spaces for exchange and dialogue across and between social levels	• Convene learning and analysis events and processes for network partners • Ground-Level Panel approach • Bring grounded learning PAR to national/global spaces	• Networked learning to identify resonances across contexts scales up evidence weight • Scale jumping by using national/global decision-making mechanism at ground level

Scaling PAR in Reality: Case Examples with *Praxis* in India

Participate steering partner Praxis is an experienced Indian NGO specialising in participatory research and community-led change, involved with two PAR projects during phase 1. The first was with sexual minorities (sex workers and LGBT communities), who are often excluded in India due to discrimination (Praxis, 2013a). The other was with people living insecurely on the street or in urban slums or "ghettos." Although dispossessed, criminalised, and highly stigmatised, these participants had self-identified as "Citymakers" during earlier PAR processes, because their work and presence contribute to creating the city, despite reaping little economic benefit from that work (Praxis, 2013b; Wheeler et al., 2020).

In this section, we reflect on our direct collaborations using participatory video to drive and mediate PAR scaling alongside other methods within these wider Praxis projects.

Nurturing Community Emergence Through Fostering Local Ownership and Inclusive Dynamics

Our Citymaker collaboration began with an eight-day participatory video process involving ten participants from three precarious living contexts (Praxis, 2013b). These were a longstanding street-dwelling site, an "objectionable" slum (meaning the authorities want to remove it), and squalid tenements that housed people forcefully relocated during slum clearance. Praxis facilitators had already run workshop sessions to orientate participants to the global UN deliberation context. Then Jackie and the Praxis team facilitated initial video exercises over two days, alongside participatory mapping, group discussion, and clustering activities. These activities aimed to prompt people to share experiences, discuss issues, and make sense of their situations together.

In broad terms, Participatory Video is a facilitated group process mediated by video recording and playback, but there is considerable methodological diversity. Using video for PAR scales down knowledge production and generates participant stories. However, Shaw (2015) has identified that short-term production processes can limit meaningful participation or lead to takeover of project benefits by the most powerful local actors. Consequently, she reconceptualises participatory video's application to PAR as an essentially relational practice, most ethically applied through extended timescales (temporal scaling). Full details of such processes are beyond this chapter's scope (Shaw, 2015, 2021), but most relevant here are the progressive video recording and playback exercises used in first stages to engage group participants.

Praxis found that excluded communities are weary of being engaged *on* rather than *with*. In response, PAR can seed community-led processes if the exploratory phase focuses on local priorities. This builds enthusiasm for action, and in turn, local ownership increases the likelihood of sustainability, in contrast to scaling out interventions developed elsewhere (Burns & Worsley, 2015). Participatory video contributes to this local focus as people are motivated by recording their own material (Shaw, 2017). Successes can then build communication confidence and counter a sense of hopelessness (Praxis, 2013b). As the Citymaker participants reported, "we women never leave our homes . . . but did not want to miss this opportunity" (Sudha, Citymaker evaluation meeting, 2013) and felt "very happy . . . all of this done by ourselves. I don't even know how to read or write" (Amulu, Citymaker evaluation meeting, 2013). However, these positive participatory video experiences did not happen automatically, and were dependent on the particular approach.

Notwithstanding the benefits of participatory video, there is a parallel risk of discomfort if not sensitively supported (Shaw, 2020). There is a danger of re-enforcing marginalisation if there is not attention to dynamics *within* the group. Thus, this participatory video stage invests time in building participant agency and generating enabling interactional contexts. One practical example is the use of a turn-taking approach to generate inclusive dynamics (Shaw, 2017). In each short exercise, participants take turns in front of the camera as presenters/ actors, and in the various videoing roles, such as camera operator, sound recorder, floor manager, and director. This approach is well evidenced for increasing a sense of "can-do" (Shaw, 2017; Wheeler et al., 2020), and a collaborative team dynamic (Shaw, 2020). However, it necessitates practitioners using their influence assertively to normalise taking turns. For example, we had to intervene to stop some men dominating the camera and ensure women's involvement (Wheeler et al., 2020). While the power dynamics that manifest during facilitated interventions may not replicate the wider structural forces that marginalise particular groups, power-aware facilitation is crucial to inclusive participation (Howard et al., 2018), and exemplifies the positive effects of participatory governance (Kesby, Kindon & Pain, 2007).

The conceptual section raised the idea of scaling long and deep to achieve meaningful participation. This case illustrates the opportunities "slower" participatory video can offer in generating the relational preconditions needed to scale deep inclusively, as well as the kinds of tensions that must be negotiated. We now consider how the process of scaling deep and out progressed in the Citymaker project.

Scaling Deep and Out Through Iterative Learning Processes

PAR participants in marginalised contexts may not realise their knowledge is valuable, and therefore rarely have preidentified priorities. Deeper knowledge on systemic issues and power-shifting solutions can be developed if there is time for iterative learning through progressive PAR cycles. Story generation was a key Participate method, because subjective narratives can reveal how macro-influences manifest at micro- and meso-levels in participant's lives. Therefore, after initial video work Citymakers next spent five days generating video stories. We facilitators used a story arc to prompt participants to first consider issues in one activity, followed by practical improvement ideas in a second (towards scaling deep). Participant subgroups, divided by their living context, then constructed a storyboard about how positive change had or could happen in their locality. Jackie, the Praxis facilitators and the Citymaker participants next spent a day recording at each site. Taking turns continued to ensure the progressive development of all participants' capacities. Practicing communicating in the different community locations was the first scaling out activity, and further prepared group members to express themselves in wider public contexts during the scaling up action that followed.

Pertinent new knowledge arose through scaling PAR out from the early project activities inside the Praxis base to the external living sites (Shaw, 2017). Contrary to the assumption that the street dwellers must be most dissatisfied, reflecting *on* the video fieldwork action afterwards suggested the reverse was true. Street dwellers face many difficulties, but are relatively in control of their lives, with regular work and strong support networks. They were also energised through acting together to secure electricity and pension provision (Praxis, 2013b). By comparison, the relocated tenement residents were extremely angry due to lost livelihoods, grossly inadequate services, rising alcoholism, and sexual abuse. These nuances of place provided important insight, identifying hopelessness and frustration as change barriers and collective action as an enabler. These insights supported the case for developing the Citymaker identity further to increase collective power, and the comparisons highlighted the need to recognise difference. We also learned *in-action* how the dynamics of PAR can change as projects scale out to involve other community members. In this instance, the Citymakers had to negotiate some tense interactions as many angry residents came out of the tenements to communicate their dissatisfaction on video, and the group's carefully crafted narrative fell by the wayside as they included this footage in the production activities.

PAR processes that reveal structural forces behind inequality can assist in targeting effective change action (Burns & Worsley, 2015), and this section began with the need for deep scaling to understand systemic issues. It also raises the challenges of scaling out small group processes whilst maintaining inclusion, as is explored further in the next section.

Navigating the Tension Between Building Collective Identities and Recognising Difference When Scaling Out

Praxis' long-term PAR with sexual minorities began as an HIV/AIDS intervention programme. The programme had been supporting community-based organisations (CBOs) in community-led monitoring, and it was apparent early on that stigma was an underlying root of HIV/AIDS, rather than a symptom (Bharadwaj, Mishra, & Raj, 2015). Following this insight, the programme's focus shifted to supporting CBOs' collective action to create safe spaces for their members, to understand cross-sectional experiences, and to engage wider stakeholders in tackling stigma. Praxis' scaling pathways involved intentionally building solidarity between diverse people living in poverty, to increase their power to leverage governance responsiveness. They suggested running a week-long participatory video project in Chennai[3] with a group of transgender participants from this wider research programme, partly because it offered potential to build collaboration between different transgender organisations in the city (scaling out) (Figure 8.1). This

Figure 8.1 Participatory video exercise with transgender participants in Chennai

project exemplified how power dynamics can become harder to negotiate as project deliberation spaces diversify during scale out.

Directly after the participatory video project, Praxis viewed the development of stronger working relationships across transgender organisations as a major achievement. However, tensions later erupted between members of the diverse organisations due to intragroup differences, such as caste and relative wealth. Nevertheless, this cross-group "storming" stage (Tuckman, 1965) ultimately proved productive, because the bonds formed between members during the participatory video process enabled the critical intersectional issues to be expressed and resolved through dialogue. Nevertheless, this incident raised the importance of addressing difference when building collective identities. This learning resonated across the Participate network, as scaling out from relatively homogenous small groups to involve wider stakeholders and leverage influence often fails to meaningfully include the least powerful groups. This learning led to a focus on addressing intersecting inequalities in the most recent Participate phase.

So far, this section focused on participatory video's contribution to intersubjective exploration and building group agency and collective purpose. Next, we consider the opportunities and dilemmas of collaborative videomaking as the means to amplify participant's perspectives when scaling out and up.

Negotiating Between Grounded Video Research Processes and Scaling Up to Influence Policy

The tension between process-focused and product-focused practice is well covered in the participatory video literature (Shaw, 2015; White, 2003) but is an unhelpful distinction, because process and video products are interconnected. Recording video material provides process motivation and direction, and playback then mediates dialogue as part of the communication context, which can stimulate further production action. The practice tensions are more clearly related to the kinds of videos (i.e., form, process of creation, technical quality, and content) that are suitable for different audiences and screening forums. Here, we focus on the scalar dimensions of these tensions.

The Participate programme aimed to first generate deep insight through long-term local PAR processes, and then scale up the knowledge and recommendations generated by bringing them into policy space. Video seemed an obvious medium to connect UN decision makers with marginalised people's realities in an emotionally affecting way (Shahrokh & Wheeler, 2014). However, ethical risks such as inappropriate exposure arise in participatory video when there is not sufficient time or attention paid to making decisions about what to screen, to whom, and how (Shaw, 2017, 2020). Shaw's (2020, 2021) extended participatory video process navigates this issue through strategies such as maintaining a clear separation between confidential recordings made to prompt internal group discussion, and videomaking focused on external audiences later in a process. This separation was used in some Participate projects. For example, in a Nairobi project, participatory video was the main PAR driver, and marginalised participants produced different videos sequentially over nine months for progressively diversifying audiences (Shaw, 2021).

However, for Participate partners such as Praxis, whose long-term PAR had predominantly used other methodologies, there was an opportunity to scale up and influence the fast-moving UN policy agenda through co-constructing a policy-focused documentary message with professional support (Robertson & Shaw, 2014). We were aware there were narrative ownership tensions created between this opportunity and participants driving their own communication agendas through participatory video. *Praxis* suggested the Chennai transgender project as context to pilot this tricky link between locally focused participatory video processes and film co-construction for the UN audience, because group members were already active leaders who would benefit from a more rapid film-making opportunity. We therefore first ran a shorter participatory video process to separate *confidential* recording for capacity building and group exploration from policy-focused production later in the weeklong project. This later phase involved participatory storyboarding, in-camera editing[4] (Shaw, 2017), and parallel production on the camcorder and professional kit to sustain group control during co-construction of the policy message (Robertson & Shaw, 2014). During evaluation, the participants reported feeling narrative ownership. Nevertheless, conflicting communication priorities emerged.

First, there was group excitement at unintentionally recording a neighbour making a derogatory expression behind one of the transgender participants while she was interviewed. Participants proposed videoing everyday discrimination and police harassment. We were uncomfortable with the idea, and alert to the risk of prompting violent responses in contested public space, which raises the issue of how to balance between practitioners' ethical boundaries and participants' control. Second, the transgender group dramatised individual stories that illustrated common discriminatory experiences, yet being close to Kollywood (the film industry in Tamil Nadu state), these scenes followed that filmmaking style and were highly dramatic, playing with illusion rather than realism. We didn't think this would translate well into UN policy space, and neither did we want to perpetuate harm by depicting overt violence. This tension illustrates Kindon's (2016) insight that participatory video or collaborative filmmaking, even when there is time to build more equal research relationships, can impose a colonial way of looking due to dominant production conventions and sensitivities. There was also a facilitatory dilemma between ethics and what was needed to scale up to communicate in UN space versus the group agendas. In this case, the lack of time prevented us from translating our knowledge of global spaces adequately.

These dilemmas were negotiated successfully due to the long-term relationships Praxis had with the groups, and because we had been transparent about the project purpose to scale up messages to the UN audience. Participate also used a share-alike copyright agreement so the partners could each use the same footage for different purposes. Two videos were produced from the recorded materials: *Marching Towards Acceptance* (Real Time, 2013a) was edited by the professional video partner Real Time to communicate the group's policy-focused call for transgender rights, and Praxis supported participants to edit another video to disseminate locally themselves (Praxis, 2013c).

Nevertheless, the interlinked and complex problems transgender participants faced in their lives were hard to communicate in-depth on video. This difficulty also arose in the Citymaker project, where Praxis supported the group to edit *Chennai – Of the Mighty and Mangled* (Praxis, 2013d). The film communicated the experience continuum across the different living contexts, in accord with the Praxis aim to build a shared Citymaker identity. However, the nuanced differences between settings were not reflected, and were thus masked by the production of a collective narrative. The global Participate network identified the need to navigate between building collective identities and addressing intersecting inequalities when scaling PAR processes (Burns et al., 2013.; Howard et al., 2018). This reflects the difficulties presented by increased heterogeneity as the scale of engagement widens and shows the importance of supporting agonistic pluralism (Mouffe, 2013) rather than false consensus during communication processes (Mistry & Shaw, 2021). To support this, the three separate Citymakers stories were all incorporated by Real Time into the Participate documentary *Work with Us*, which was aimed at national and global policy space (Real Time, 2013b).

Video offers potential for communicating ground-level learning from PAR processes, but we found that the inherent tensions are amplified and made more complex as the engagement level moves outwards and upward. This emphasises the need to carefully consider video form, quality, content, and the distribution mechanism in relation to the scale of communication aims and the audience's spatial positioning. Most crucially, we approached video not as the end point of PAR processes, but to generate further conversations as the processes scaled out and up, as we discuss next.

Scaling Out and Up Through Participatory Deliberation and Community-Led Action

Input from IDS to network partners aimed to extend the scaling potential of the Participate PAR processes through using visual methodologies, as did sharing methodological learning between partners. To support methodological quality and embed new approaches, Jackie delivered participatory video training for Praxis staff and participants in India, who also attended Participate digital storytelling training in the UK. This training was scaling in to nurture methodological uptake. Praxis then scaled out by conducting further participatory video processes, digital storytelling, and mapping activities. The original Citymakers and transgender participants were from Chennai, but horizontal scaling moved to different urban areas to recruit participants with similar socioeconomic backgrounds. For example, in Delhi new Citymakers and sewage workers were recruited in four areas, and sex workers participated in Maharashtra and Tamil Nadu. Participants analysed their experiences themselves, fostering knowledge ownership. Group recommendations were published as Voice for Change reports (e.g. Praxis 2013a, 2013b), and participants used these outputs and their videos to facilitate dialogue with external audiences and support their own agendas (Narayanan et al., 2015). For example, the sexual minorities network wanted to increase their policy visibility, alongside provision of drop-in safe spaces and crisis responses. They screened their film to the Transgender Welfare Board in Tamil Nadu, as well as to NGO employees, transgender and sexual minority CBOs, five colleges in Chennai, and to the public, homeless people, and in slums. The film was also selected for presentation at the VIBGYOR International Film Festival in Kerala. The screenings were impactful. For example, a government officer from the Tamil Nadu welfare board expedited free housing allocation to transgender recipients after watching the film. *Down the Drain* was produced by Delhi sewerage workers about their appalling work conditions and lack of safety. When it was shown during a Supreme Court hearing on contract workers' rights, it triggered progressive discussion about their urgent protection needs. Critically, it was because these scaling out processes were community-driven that they were able to gain momentum and sustain advocacy action.

The development of international policy initiatives like the SDGs is normally conceived as something to which only national or international experts might be expected to contribute. Yet this conventional scale for policy formulation can be questioned in relationship to the long-established PAR principal that ordinary people are experts in their realities, and that as the SDGs will shape their everyday lives, they should be involved in formulating and monitoring them.

Scale jumping involves reconceptualising the scale at which a process operates. Praxis applied these scalar politics when it developed what were termed Ground-Level Panels (GLP) (Narayan et al., 2015) to mirror the "expert" High-Level Panel (HLP) of UN policymakers. GLPs were national forums, but also grounded because panel members had lived experience of poverty and marginalisation and spoke about the realities for people like them across the nation. The Indian GLP brought 14 diverse people together in a five-day process to compare their experiences, reflect on videos and other evidence, and systematically analyse the gaps in the UN goals. For example, they suggested the goal on ending poverty required action on corruption. Reflecting the panel's composition, they emphasised that focusing on women's empowerment neglected gender equality for transgender people. Participants then presented their conclusions publicly to an audience of national government officials, development organisations, NGOs, the media, and groups like themselves who had been excluded from these processes. This GLP approach was later adapted for Participate partners in Brazil, Egypt, and Uganda.

Scaling Up Through Bringing PAR-Generated Learning into Global Policy Spaces

Participate set out to foster the conditions for wider social impact from local PAR processes, by instigating the network and facilitating collective learning. For example, a four-day global synthesis workshop involved partners in analysing resonances and differences from the diverse PAR processes across 30 countries. Triangulating these local insights transformed them into collective global evidence, making research conclusions more convincing for international audiences. IDS then brought this research knowledge into formal UN policy processes to optimise our collective influence (scaling up). One partner, Beyond 2015, contributed by identifying "opportunity moments" in the unfolding global dialogue (Shahrokh & Wheeler, 2014), which enabled Participate to input throughout the SDG deliberation process. For example, the transgender film was screened to a global High-Level Panel Forum in Bali. Finally, the videos alongside visual outputs from the 30 Participate countries formed part of physical and online exhibitions before the UN SDG summit in New York (Shahrokh & Wheeler, 2014). There were also a series of engagement events, including the opportunity for policymakers to meet participants virtually after watching their videos.

We were told anecdotally that Participate critically influenced formation of the *leave-no-one behind* narrative at these summits (Burns & Worlsey, 2015). We attributed this in part to the use of visual methodologies. Media such as video used scale in innovative ways – it enabled marginalised participants to be seen and heard at global level, located visually in their everyday environments, and speaking convincingly in their own words. This engaged audience attention in a human-to-human and emotionally compelling way, with some global actors moved to support the Participate programme as long-term allies.

Notwithstanding these benefits, there remains a risk of miscommunication and audience misunderstanding (Shaw, 2017). For example, the IDS team experienced communication gaps across levels. At the global level, the SDG process was structured around discrete goals, and it was clear in our early discussions with funders (DFID) that they wanted us to tell them which services participants prioritised. Meanwhile, a key overall finding from the combined PAR projects was that participants themselves wanted a change in *how* development is *done* (Burns & Worsley, 2015). Across country contexts, we repeatedly heard that there were already health, education, and justice services, but our participants were unable to access them due to discriminatory attitudes.

Additionally, it was clear from our work that the effectiveness (or ineffectiveness) of policies lay in how they were implemented, monitored, and adapted in situ. Consequently, the two main messages that Participate transmitted to policymakers were: a) that marginalised people want to be involved in making change happen, and b) that they *cannot tackle wider structural social problems alone* (Real Time, 2013b). Their involvement in change-making action requires support and resourcing, and they wanted decision makers to work *with* them on community-driven change processes, rather than impose programmes top-down.

However, the network identified that a key sticking point on pathways to scale lies in holding duty bearers to their commitments to community-led action. Hence the Participate programme next explored how to foster accountable relationships between people facing intersecting inequalities and influential external allies.

Persevering Through the Setbacks with Hope

In the latest Participate phase (2016 to 2018), Praxis collaborated with "De-notified and Nomadic Tribes" (DNTs) (Naryanan, Dheeraj, & Bharadwaj, 2020; Shaw et al., 2020). Approximately 198 DNTs were identified as highly "left-behind" communities because they were labelled as criminal from birth during the colonial era, and continue to face extreme stigma.

The PAR processes Praxis conducted with DNT communities built on the previous experiences discussed above (Naryanan et al., 2020). This time, the strategy was to increase scaling opportunities by aligning the PAR project with a national DNT campaign, and codesigned the process. Peer-led research with 174 respondents across five states used household surveys, focus group discussions, and digital storytelling. From amongst these participants, 11 people from diverse tribal groups took part in a ground-level panel and then presented their recommendations alongside a short video at a national policy event and a UN high-level political forum discussion. As with previous projects, Praxis aimed to build a collective DNT identity to increase political leverage. However, this proved challenging due to the diversity of tribal groups, another manifestation of scale complexity. At a local scale, a group can identify mutual experiences and forge a common purpose, but scaling up requires that their interests and identities be amalgamated into a new shared category at a wider scale (e.g., DNT people). This can cause difficulties because participants may not want to be seen as similar to each other, at least at first. For instance, women in Nat, Bedia, or Kanjar communities were traditionally often involved in sex work, and other DNT groups did not want to associate with them, which was a barrier to embracing a DNT identity (Howard et al., 2018). Praxis found it hard to ensure women's inclusion in general, because the effects of patriarchy meant men tended to dominate. Finally, they found that despite the strength of participants' video stories, deep-seated prejudices among audiences meant that viewers tended to respond moralistically to predicaments represented in the videos (Naryanan et al., 2020). These examples show the challenges of increased heterogeneity as scalar units increase in size. In addition to these issues, the government department responsible for DNTs closed during the research. This had the effect of undermining the legitimacy and utility of the DNT scale unit the project was attempting to mobilise.

While these significant obstacles cannot be dismissed, it is also true that change does not happen in predictable ways. Small wins can be followed by setbacks, or vice versa, and persevering can help create the conditions for an eventual tipping point (Green,

2016). Just recently, the work to mainstream DNT communities has paid off. The Indian government and UN agencies have identified DNT as one of seven left-behind or most excluded groupings and propose engaging with these communities through the research group and *Praxis*. This has revitalised DNTs as a politically relevant scale unit and given new impetus to find ways to work across differences among heterogenous groups with many (but not all) interests and experiences in common. Onwards . . .

Conclusion: Navigating the Tensions on the Pathways to Scale Through PAR

Scaling down PAR processes to include the most marginalised people in context, and scaling up the insights and socially inclusive practices generated to maximise improvements, are persistent challenges at the cutting edge of inclusive development practice. The Participate programme provided a unique opportunity to explore how to catalyse largescale pathways to social impact from the local to global and back again. Drawing on our direct experiences using participatory video methodologies as part of Praxis' long-term PAR processes, and in relationship to the wider Participate programme, we have reflected in this chapter on how to navigate the pathways to scale from the social margins. We do not wish to suggest that participatory video or collaborative filmmaking were the significant activities. Rather, we want to illustrate the possibilities and tensions in using them within wider research and community mobilisation processes, to generate inclusive communicative space, build shared purpose, increase group agency, and to amplify communication links when scaling out and up (Shaw, 2015, 2021).

Participate applied recent thinking on scaling social inclusion innovation, which suggests its dynamics are like nurturing a social movement. We considered different stages of three Praxis projects to show how Participate's main scaling strategies unfolded in practice. Again, the intention is not to propose one right way to scale down, out and up through PAR. Rather, approaches should be contextualised and adapted responsively as a process evolves. That requires ongoing reflection, power-aware collaboration, and iterative learning cycles. PAR processes generally operate in contested contexts, and always encounter unavoidable tensions along the way. Previous research also guided our understanding that visual methodologies can increase both possibilities and risks (Shaw, 2017, 2020; Wheeler et al., 2020). Traveling the pathways to scale through PAR therefore involves ongoing navigation to maximise opportunities, given the linked challenges. We used the lenses of scale and scaling action to illuminate what this meant in context.

Deleuze and Guattari (1987) clarify that life does not exist in neat conceptual categories. As highlighted at the beginning of this chapter, theory is only useful if it productively illuminates the nuances *between* social constructs. Were the scaling concepts productive, or would a flat ontological lens better guide transformative action? In one sense, Participate worked with hierarchical scalar imaginaries, particularly as these constructed what was possible in the SDG process. At the same time, the inescapable scaler organisation of UN processes provided entry points for disrupting expectations by making new connections across and between levels.

Reflecting on our experiences through a scaler lens enabled us to see the way Participate linked temporally long but local PAR processes. It supported them to scale both deep and out and scaled up the collective knowledge generated to influence global policy. Scale concepts also highlighted how power dynamics become harder to navigate as deliberative spaces increase in size and project actors diversify. They also demonstrated the danger of perpetuating inclusion when mobilising collective action. We have recommended ways to negotiate these challenges throughout the chapter. Action at "higher" social levels or with increased participant numbers or diversity requires time for iterative processes and careful facilitation, as well as even more attention to building relationships. However, there are no ideal projects, and sometimes Participate partners had to move very quickly to meet a policy deadline. We were able to do so because speedy responses were based in the previous work to establish trusting relationships.

What became increasingly clear is that what most needs to be scaled are approaches to generating inclusion through shifting power dynamics. Despite the challenges, PAR offers this potential. Emergent local research processes can reach and generate knowledge about places, people, and situations because they are not pre-conceived or directed from outside. These processes can drive innovation, and inclusive local ownership can sustain action as processes scale out. As anticipated, tensions were intrinsic when scaling up and out, but PAR made tacit constraints more *visible*. PAR made it more possible to anticipate the worst risks and mitigate or respond to them in situ. This knowledge about navigating tensions supports inclusiveness in future scaling projects.

Notes

1 The Participate research initiative was co-convened by the Participation, Inclusion and Social Change Cluster at the Institute of Development Studies (IDS) and the civic society campaign *Beyond 2015*. See https://www.ids.ac.uk/projects/participate-knowledge-from-the-margins-for-post-2015/.
2 See https://www.praxisindia.org/.
3 Chennai is the capital of Tamil Nadu state in South India.
4 In-camera editing involves recording a sequence of video shots one-by-one so that no computer editing is required afterwards.

Acknowledgments

We thank Mike Kesby for his helpful reviewing feedback and suggestions, which provoked deeper reflection and contributed to a stronger chapter. We acknowledge the considerable input from the collaborating partners in the Participate network and the many project participants globally, which has contributed to the richness of programme learning and therefore our insights.

References

Amin, A. (2002). Spatialities of globalisation. *Environment and Planning A: Economy and Space*, 34 (3): 385–399. https://doi.org/10.1068/a3439.

Bharadwaj, S., Mishara, S., & Raj, A. M. (2015). Subverting for good: Sex workers and stigma. In T. Thomas & P. Narayanan (Eds.), *Participation Pays*. Rugby: Practical Action Publishing.

Brenner, N. (2001). The limits to scale? Methodological reflections on scalar structuration. *Progress in Human Geography*, 25(4): 591–614. https://doi.org/10.1191/030913201682688959.

Burns, D., Howard, J., López-Franco, E., Sharokh, T., & Wheeler, J. (2013). *Work with Us: How People and Organisations Can Catalyse Sustainable Change*. Brighton: IDS.

Burns, D., & Worsley, S. (2015). *Navigating Complexity in International Development*. Rugby: Practical Action Publishing.

Chambers, R. (1997). *Whose Reality Counts? Putting the First Last.* Rugby: Practical Action Publishing.

Chambers, R. (2005). *Ideas for Development.* London: Earthscan.

Cohen, A., & McCarthy, J. (2015). Reviewing rescaling: Strengthening the case for environmental considerations. *Progress in Human Geography*, 39(1): 3–25. https://doi.org/10.1177/0309132514521483.

Cooke, B., & Kothari, U. (2001). *Participation: The New Tyranny?* London: Zed Books.

Darian-Smith, E., & McCarty, P. (2017). *The Global Turn: Theories, Research Designs and Methods for Global Studies.* Berkeley: University of California Press.

Deleuze, G., & Guattari, F. (1987). *A Thousand Plateaus: Capitalism and Schizophrenia.* New York: Continuum International.

Dinham, A. (2005). Empowered or over-powered? The real experiences of local participation in the UK's New Deal for Communities. *Community Development Journal*, 40(3): 301–312.

Escobar, A. (2007). The 'ontological turn' in social theory. A Commentary on 'Human geography without scale.' *Transactions of the Institute of British Geographers*, 32: 106–111. https://doi.org/10.1111/j.1475-5661.2007.00243.x.

Foucault, M. (1980). *Power/Knowledge: Selected Interviews and Other Writings, 1972–1977.* Brighton: Harvester.

Fox, J. (2015). Social accountability: What does the evidence really say? *World Development* 72, 346–361.

Gaventa, J., & Cornwall, A. (2008). *Power and knowledge.* In P. Reason & H. Bradbury (Eds.), *The Sage Handbook of Action Research: Participative Inquiry and Practice.* 172–189. London: Sage Publications.

George, G., McGahan, A. M., & Jaideep, P. (2012). Innovation for inclusive growth: Towards a theoretical framework and a research agenda. *Journal of Management Studies*, 49(4): 661–683. https://ink.library.smu.edu.sg/lkcsb_research/4674.

Green, D. (2016). *How Change Happens.* Oxford: Oxford University Press.

Herbert, S. (2005). The trapdoor of community. *Annals of the Association of American Geographers* 95(4): 850–865.

Howard, J., López Franco, E., & Shaw, J. (2018). *Navigating the Pathways from Exclusion to Accountability: From Understanding Intersecting Inequalities to Building Accountable Relationships.* Brighton: IDS.

Howard, J., & Shaw, J. (2024). *Building sustainable inclusion: From intersecting inequalities to accountable relationships.* Institute of Development Studies. https://www.ids.ac.uk/projects/building-sustainable-inclusion-from-intersecting-inequalities-to-accountable-relationships/

Jones, K. (1998). Scale as epistemology. *Political Geography*, 17: 25–28.

Kabeer, N. (2016). "Leaving no one behind": The challenge of intersecting inequalities. In International Social Science Council and Institute of Development Studies (Eds.), *World Social Science Report 2016: Challenging Inequalities: Pathways to a Just World.* Paris: UNESCO Publishing.

Kemmis, S., & McTaggart, R. (2005). Participatory action research: Communicative action and the public sphere. In N. K. Denzin & Y. S. Lincoln (Eds.), *The Sage Handbook of Qualitative Research* (3rd ed). 559–604. Thousand Oaks, London, New Delhi: Sage Publications.

Kindon, S. (2016). Participatory video as a feminist practice of looking: Take Two! *Area*, 48: 496–503. https://doi.org/10.1111/area.12246.

Kesby, M. (2007). Spatialising participatory approaches: The contribution of geography to a mature debate. *Environment and Planning A*, 39(12): 2813–2831. https://doi.org/10.1068/a38326.

Kesby, M., Kindon, S., & Pain, R. (2007). Participation as a form of power: Retheorising empowerment and spatialising Participatory Action Research. In S. Kindon, R. Pain, & M. Kesby (Eds.) (2007), *Participatory Action Research Approaches and Methods: Connecting People, Participation and Place.* 19–25. Abingdon: Routledge.

Kindon, S., Pain, R., & Kesby, M. (Eds.) (2007). *Participatory Action Research Approaches and Methods: Connecting People, Participation and Place.* Abingdon: Routledge.

Ledwith, M., & Springett, J. (2010). *Participatory Practice: Community-Based Action for Transformative Change.* Bristol: Policy Press.

Leitner, H., & Miller, B. (2007). Scale and the limitations of ontological debate: A commentary on Marston, Jones and Woodward. *Transactions of the Institute of British Geographers*, 32: 116–125. https://doi.org/10.1111/j.1475-5661.2007.00236.x.

López Franco, E., & Shahrokh, T. (2024). *Participate: Knowledge from the margins for post-2015.* Institute of Development Studies. https://www.ids.ac.uk/projects/participate-knowledge-from-the-margins-for-post-2015/

Marston, S. A., Jones, J. P. III, & Woodward, K. (2005). Human geography without scale. *Transactions of the Institute of British Geographers*, 30: 416–432. https://doi.org/10.1111/j.1475-5661.2005.00180.x.

McLean, R., & Gargani, J. (2019). *Scaling Impact: Innovation for the Public Good.* Abingdon: Routledge

Mansuri, G., & Rao, V. (2004). Community-based and -driven development: A critical review. *World Bank Research Observer*, 19(1): 1–39.

Maurer, M., & Githens, R. P. (2009). Toward a reframing of action research for human resource and organisation: Moving beyond problem solving toward dialogue. *Action Research*, 8(3): 267–292.

Mistry, J., & Shaw, J. (2021). Evolving social and political dialogue through participatory video processes. *Progress in Development Studies*, 21(2): 196–213. https://doi.org/10.1177/14649934211016725.

Mouffe, C. (2013). *Agonistics: Thinking the World Politically.* London: Verso.

Narayanan, P., Bharadwaj, S., & Chandrasekharan, A. (2015). Reimagining development: Marginalised people and the post-2015 agenda. In T. Thomas & P. Narayanan (Eds.), *Participation Pays.* Rugby: Practical Action Publishing.

Naryanan, P., Dheeraj, Sinha, M., & Bharadwaj, S. (2020). Participation, social accountability and intersecting inequalities: Challenges for interventions to build collective identity with De-notified, Nomadic and Semi-Nomadic Tribal communities in India. *Community Development Journal*, 55 (1): 64–82. https://doi.org/10.1093/cdj/bsz034.

Park, P., Brydon-Miller, M., Hall, B., & Jackson, T. (Eds.) (1993). *Voices of Change: Participatory Research in the United States and Canada.* Westport: OISE Press.

Praxis (2013a). *Voice for Change: Collective Action for Safe Spaces by Sex Workers and Sexual Minorities.* Delhi: Praxis.

Praxis (2013b). *Voice for Change: Citymakers Seeking to Reclaim Cities They Build.* Delhi: Praxis.

Praxis (2013c). *Beyond 2015 – Voices for Empowerment.* YouTube. https://www.youtube.com/watch?v=O-oMKwKlX38&t=260s

Praxis (2013d). *Chennai – Of the Mighty and Mangled.* Youtube. https://www.youtube.com/watch?v=4iU5C9CGXog&t=4s

Rademacher, A., & Patel, R. (2002). Retelling worlds of poverty: Reflections on transforming participatory research for a global narrative. In K. Brock & R. McGee (Eds.), *Knowing Poverty: Critical Reflections on Participatory Research and Policy.* 166–188. Oxford: Earthscan.

Real Time (2013a). *Marching Towards Acceptance.* Vimeo. https://vimeo.com/74171698

Real Time (2013b). *Work with Us.* Vimeo. https://vimeo.com/showcase/4488354/video/80075380

Reason, P., & Bradbury, H. (Eds.) (2008). *The Sage Handbook of Action Research: Participative Practice.* London: Sage Publications.

Robertson, C., & Shaw, J. (2014). Balancing ground level visual research processes and communication products for policy influencing. In T. Shahrokh & J. Wheeler (Eds.), *Knowledge from the Margins: An Anthology from a Global Network on Participatory Practice and Influence.* 39–40 & 54–55. Brighton: IDS.

Sassen, S. (2008). *Territory, Authority, Medieval to Global Assemblages.* Princeton, NJ: Princetown University Press.

Shahrokh, T., & Wheeler, J. (Eds.) (2014). *Knowledge from the Margins: An Anthology from a Global Network on Practice and Policy Influence.* Brighton: IDS.

Shaw, J. (2021). Extended participatory video research processes. In D. Burns, J. Howard, & S. Ospina (Eds.), *The SAGE Handbook of Participatory Research and Enquiry*. London: Sage Publications.

Shaw, J. (2020). Navigating the necessary risks and emergent ethics of using visual methods with marginalised people. In S. Dodd (Ed.), *Ethics and Integrity in Visual Research Methods*. 105–130. Emerald Publishing Limited. https://doi.org/10.1108/S2398-601820200000005011.

Shaw, J. (2017). Pathways to accountability from the margins: Reflections on participatory video practice. *Making All Voices Count Research Report*. Brighton: IDS.

Shaw, J. (2015). Re-grounding Participatory Video within community emergence towards social accountability. *Community Development Journal*, 50(4): 624–643. https://doi.org/10.1093/cdj/bsv031.

Shaw, J., Howard, J., & López Franco, E. (2020). Building inclusive community activism and accountable relations through an intersecting inequalities approach. *Community Development Journal*, 55(1): 7–25. https://doi.org/10.1093/cdj/bsz033.

Smith, N. (1992). Geography, difference and the politics of scale. In J. Doherty, E. Graham, & M. Malek (Eds.), *Postmodernism and the Social Sciences*. London: Palgrave Macmillan. https://doi.org/10.1007/978-1-349-22183-7_4

Stein, C., & Jaspersen, L. J. (2021). Participatory network research: Using visual methods and participatory statistics for value chain analysis. In D. Burns, J. Howard, & S. M. Ospina (Eds.), *The SAGE Handbook of Participatory Research and Enquiry*, London: Sage Publications.

Tuckman, B. W. (1965). Development sequence in small groups. *Psychological Bulletin*, 63(6): 384–399.

Watt, D. (1991). Interrogating 'community': Social welfare versus cultural democracy. In V. Binns (Ed.), *Community and the Arts: History, Theory, Practice*. London: Pluto Press.

Wheeler, J., Shaw, J., & Howard, J. (2020). Politics and practices of inclusion: Intersectional participatory action research. *Community Development Journal*, 55(1): 45–63. https://doi.org/10.1093/cdj/bsz036.

White, S. A. (2003). *Participatory Video: Images that Transform and Empower*. New Delhi, London: Sage Publications.

9 Movement Memories in the Afterlife of Participatory Action Research (PAR)

Dreaming and Forgiveness Beyond the Non-Profit Industrial Complex (NPIC)?

Amy Ritterbusch

Introduction

My dear sister-in-struggle[1] (*hermana de lucha*) and street-level activist Daniela Maldonado Salamanca presented the following discourse denouncing "Vanilla Feminist" extractive research, activism, and institutionalised social justice practices at a protest in the streets of the sex work zone in Bogotá:

> Today is the 8th of March. Not like any other day. Today, as the nobodies of feminism, WE ARE FURIOUS! . . . Our rage overcame their good manners and their academic arguments. They are not enough for us! THEIR VANILLA FEMINISM EXCLUDES US! Today is the 8th of March where we, the ever-violated ones, are tired of being just that: victims who increase statistics in human rights reports. NEVER AGAIN! WE ARE WARRIORS, FIGHTERS, INVINCIBLE, IMMORTAL!
>
> Today is an 8th of March that changes history because we, the "never nobody" women, scream from the corners where we work.
>
> From the high heels that empower us, from the streets through which we march.
>
> We demand our rights! ([echo] – rights – rights – rights). . . . Because we are not desk lesbians, nor are we NGO lesbians; we are the real rug munchers, the ones of the people's class, the ones of the neighbourhoods, of the plazas!
>
> (Maldonado Salamanca, 2019[2])

Daniela's manifesto gives me chills every time I share it – in classroom spaces, in processes of collective writing and reflection, and in our *Sueños Furiosos* volume we collectively published recently (Ritterbusch & González, 2019). There is so much wisdom in her lessons on justice-seeking. This manifesto and so many other artistic and protest-immersed interventions merit a deep process of unpacking and collective learning for solidarity activism and research around the world. Here, I focus on two main arguments in her message to the academy and to non-profit-driven activism.

First, she notes rage. Daniela's rage here is directed against those bookish and theoretical academics and activists who have excluded her sisters-in-struggle from certain (white and "well-behaved") feminisms, and from academic discourses that intentionally obscure and exclude their voices. "They don't deserve us!," she proclaims as she alludes to these exclusions of her community, and their simultaneous unrelenting inclusion in human rights documents that report the running death toll and violence against trans street-level activists, sex workers and unhoused communities (as well as the connections

DOI: 10.4324/9780429400346-9

and intersections between them). Daniela's words validate the work PAR justice-seekers have insisted upon as we fight and learn with communities in the streets and bring it to our classrooms to shake up the academy.

Second, she refers to class. A class-conscious analytical lens is crucial for our imagining of other worlds that have overthrown the reign of capitalism. Her references to "desk" and "NGO" are critiques of the academics and the NGO activists who stay at their desks and have little contact with the communities they write reports about. Daniela's critique aligns with two decades of PAR theorising on the importance of participation and attention to positionalities within social movements (Askins 2018; Cahill, 2010; Fals Borda, 1991; Fine & Torre, 2006; Stoudt, Fox, & Fine, 2012; Torre, 2009; Torre & Ayala, 2009). Daniela's provocations are instructive for how scholars think about and critique the troubling entanglement of social movements with what movement scholars have referred to as the Non-Profit Industrial Complex (NPIC) (Incite!, 2007). Rodriguez (2007) defines the NPIC as "the set of symbiotic relationships that link together political and financial technologies of state and owning-class proctorship and surveillance over public political intercourse, including and especially emergent progressive and leftist social movements, since about the mid-1970s" (pp. 21–22).

Is activist-scholarship so deeply intertwined within the world of the NPIC and the Academic Industrial Complex (AIC) that other worlds beyond these extractive logics are impossible to imagine? What would worlds of activist-scholarship look like in direct opposition to these logics? It seems to me that postcapitalist research and activism, and a postcapitalist activist-scholarship, are only possible in a scenario where the AIC, or the commercialisation of research that depends on the extractive rampaging of the poor and marginalised, and the NPIC, or the commercialisation of activism and social justice advocacy work that depends on the exploitation of the poor and marginalised, are toppled.

The Incite! collection clearly sets forth the argument that this world, and its associated revolution, will (and should) NOT be funded, and that to sustain our movements toward justice, we must fiercely reject and diminish the possibility of the capitalist co-optation of our work (Incite!, 2007). It is disheartening to learn of the accusations of ethnic fraud against a member of this collective; however, violent, deceitful reproductions of coloniality are well-known patterns within the web of the AIC, as authors ironically discuss in this volume (Incite!, 2007). When our movement called PARCES (Peers in Action-Reaction Against Social Exclusion) made the decision to close our doors in 2017, we chose, however painful it was at the time, to reject and remove ourselves from this toxic web of capitalist co-optation.

J. K. Gibson-Graham (2006) suggest that a postcapitalist politics implies, "creating the world of possibility that enables 'other worlds' to actually arise" (p. xii). In this chapter I dream forward, as I remember and honour the contributions of two sisters-in-struggle who survived the rampage of NPIC/AIC logic. For me, dreaming forward is something like the poetics of struggle set forth by Robin D. G. Kelley in *Freedom Dreams* (2002):

> In the **poetics of struggle** and lived experience, in the utterances of ordinary folk, in the cultural products of social movements, in the reflections of activists, we discover the many different cognitive maps of the future, of the world not yet born. Recovering the poetry of social movements, however, particularly the poetry that dreams of a new world, is not such an easy task . . . what we are against tends to take precedence over what we are for, which is always a more complicated and ambiguous

> matter. It is a testament to the legacies of oppression that opposition is so frequently contained, or that efforts to find "free spaces" for articulating or even realising our dreams are so rare or marginalised.
>
> (pp. 9–10)

In this sense of the poetics of struggle, dreaming forward consists of doing everything possible to repel NPIC and AIC logic in my daily life. It means carving out spaces every day to articulate my dreams for a different world, and then acting upon these dreams. Teaching against the forces of NPIC and AIC in the classroom, in every paper I write, in every proposal I attempt to think through collectively with movement family. As Kelley (2002) notes, dreaming forward means dwelling in the reflections of activists, scholars, and all justice-seekers working from different spaces of struggle to articulate the contours of a world where it is possible to fight forward and fuel our movement for justice without succumbing to the web of the NPIC and AIC.

I present the ruptures that occurred within street-level social movements I have witnessed over the years and attribute these largely to the increased entanglement of community activists with the NPIC, and the subsequent frenzy of grassroots resource (mis) management. After experiencing worlds of heartbreak in my work as an activist-scholar in Colombia, I reflect on the possibilities of a postcapitalist activist-scholarship as I describe two relationships developed within a framework of participatory action research (PAR). I also present what I dream and hope for our friendships in the future. My reflections on PAR in this chapter and elsewhere centre on relationships and the deep love I hold for my sisters. I would like to make this clear at the outset. It remains unclear whether PAR lies within or beyond the NPIC and the AIC. I think we dreamed beyond both forces of oppression in our PAR praxis; however, we may never achieve complete disentanglement in our lifetimes.

These dreams look at the afterlife of organisations that crumbled under the combined pressures of state violence and capitalist extractivism on one side, and on the other, radical politics that denounced the criminalisation of poverty yet almost always reproduced capitalist and NPIC logic as they seeped into the movement's social fabric (see Incite!, 2007). In this chapter I present memories from our movement in the context of two friendships I developed in the streets, with the intention of exploring the broader entanglements of PAR with the NPIC and the AIC.

The historical emergence of PAR, reaching toward existence beyond the NPIC, emerged from social movements in Colombia, including the work of the late sociologist Orlando Fals Borda (1991, 2003) and his comrade Camilo Torres. I understand my relationships and deep love for my sisters-in-struggle as historically situated in this deep friendship, love, and troubled comradery modelled in the relationship between Orlando and Camilo. This relationship at the heart of justice-seeking is also modelled by Gibson and Graham's collective publishing practice. The editors and various contributors to this volume have also engaged in divesting from individualistic dissemination practices through collective writing practices (mrs kinspaisby-hill, 2011).

In writing this chapter, as I attempt to disentangle my world from that of the NPIC and AIC, I struggle with how to narrate these movement memories in terms that both uphold my love and respect for my sisters, and also instruct our movement and movements in different parts of the world. PAR scholars have long discussed the lack of alignment between institutional ethics and care ethics (Cahill, 2007; Cahill, Sultana &

Pain 2007; Sultana, 2007). The act of writing this chapter demonstrates the friction between these two realms of ethical reflection and action.

This piece is one of many texts I dream of writing in honour of the knowledge held by my sisters and the important lessons I have learned from them in shared spaces; in the streets, in the university, and in social justice organisational spaces laden with AIC and NPIC dynamics. Here I honour the contributions of two particular sisters-in-struggle, referred to here as "La Madre" (The Mother) and "La Tango" (The Dancer), names they assigned themselves. I refuse to give up on my relationships with them, and it has caused me great sadness to watch other activists and funders do so. Both activists have experienced exclusion from particular factions of their movement families in different moments, in large part because they fell into the quicksand constantly forming at the nexus of the NPIC, "community-based organisations," and poverty. These activists have managed resources for their communities for years, while themselves experiencing hunger and unhoused life. Both activists have been accused by different members of the community of (mis)using resources for their own benefit.

One of the central questions in this chapter is: How radical is it to ostracise members of your movement family? Where should the line be drawn between exercising accountability and condoning "resource mismanagement" and other NPIC-invoked "transgressions"? I am deeply saddened as I witness some sisters watch from afar while ostracised activists stuck in the NPIC gasp for their last breath before going under. They watch from afar as their sisters are immobilised in the abyss of the NPIC and critique their sisters' lack of movement. While in-fighting and character assassination are sadly part of the NPIC toxic web, there is something about these two sisters-in-struggle that keeps me awake at night. In this chapter, I attempt to gain clarity about what exactly disturbs my sleep and what it means for postcapitalist forward-dreaming in my PAR work and my relationships with sisters-in-struggle.

I met La Madre and La Tango at different moments in the beginnings of my activist work in the Santa Fe neighbourhood of Bogotá. Both activists were social leaders doing street-level "outreach" work in organisations I had worked with before. I have learned so much from both dear friends and sisters-in-struggle in shared activist spaces, including NGO movements, street-level protest, organising in the sex work and drug consumption zones in Bogotá, and in classroom spaces when I worked as a university professor in Bogotá.

I have argued elsewhere for the importance of keeping relationships at the centre of justice-seeking endeavours (Ritterbusch, 2019). Here I communicate my deepening heartbreak as I witness relationships torn apart by mistrust and individualism injected into street-level movement space by the capitalist ethos of the NPIC.

In this chapter, I dream of other worlds where relationships for justice are not torn apart by street-level activists' entanglements with the NPIC. I contextualise my friendship and admiration of these two street-level activists. I argue that not only are their efforts for justice-seeking stunted and '*echados al olvido*'[3], but they too are ostracised by their movement family. My observations seek to reach beyond classic accounts of in-fighting amongst activists over funding sources and beyond the well-documented accounts of activist burnout (Pigni, 2016). It is not so much about conflict over resources but how the activism running through the veins of my sisters is smothered by street survival (surviving day-by-day amid violence, political persecution and structural poverty) and by NPIC entanglement. My reflection here centres on the lack of solidarity and mutual understanding that occurs when people momentarily give up on activism, on their relationships and responsibility to the people they organise for and with and mismanage movement resources for survival. Survival mismanagement may even mean drinking or smoking the movements' collective money away as a means of numbing out personal pain and trauma, spending the money on funeral expenses, using resources to dignify lives that have been dehumanised for decades. And as in the case of both activists, within the precarity of multidimensional poverty and socioeconomic and political exclusion.

My reflections here draw from my memory of ruptures in the street-level movements I hold close to my heart. I focus on movement ruptures because I believe that learning from these conflicts may be the key for extracting ourselves from the reach of the NPIC and AIC and is necessary to dream/imagine postcapitalist research and activism. While I dream and work toward these alternatives to the current logic of justice-seeking, I also explain the current nightmare of entanglement in the NPIC and AIC as a means of processing and healing.

Both La Madre and La Tango have worked for countless organisations and research projects as community researchers and community organisers. They became entangled with the AIC and the NPIC through their years of working with these organisations as they transitioned between NGO and academic spaces. Before I explore each case in depth and how their entanglements overwhelmed their commitments to the collective, I would like to honour my relationship with both activists and share my memories of their important work for the movement.

Movement Memories of La Madre: Fierce Maestra, Defender of the Right to the City

I first met La Madre in 2008 at Procrear Foundation. She worked as a "peer leader"[4] in multiple activist projects with street-connected women and drug users in the Santa Fe area of Bogotá and survived on the small income she earned in these projects. I was a doctoral student working as a volunteer with Procrear and Fenix Social Foundations. During this time, I began to form close friendships with multiple activists in the city's street-level movements. These relationships started with my participation in a trans women's dance group. One dancer was our sister-in-struggle, Wanda Fox. Soon after meeting her, she was assassinated by paramilitary actors in the streets of Santa Fe. Wanda Fox is a legend in the street-level trans activist movement in Bogotá.

Presently, I assert that La Madre is also a legend. A day may come when, like Wanda Fox, La Madre does not return from the streets and is lost to transphobic or state violence. So, without romanticising her complex positionality, and notwithstanding her strained relationship with some members of the street-level trans movement in the city, I want to speak about her positive contributions to justice-seeking for trans women in the streets.

La Madre's smile and laughter brighten any space, any unbearable silence, any classroom. For two consecutive years, I had the honour of coteaching an urban justice course called "The Right to the City" with her at Universidad de los Andes (Uniandes), a private university in Bogotá. La Madre's contribution to the teaching program was sustained by an Open Society Foundations grant. Vital though this was, it implicated us in the NPIC.

Instead of holding our sessions in the class-privileged space of Uniandes, amid security guards and luxurious cafes, our learning space was based in the headquarters of the Trans Community Network in Santa Fe. Santa Fe is the sex work zone in Bogotá and a space where other street-connected communities seek refuge, including street vendors, unhoused individuals, and drug users.

Some of my fondest pedagogical memories are in this classroom, which radically subverted hierarchies of knowledge production and the exclusionary history of what counts as a legitimate trajectory of professorship. Our classroom was the street. In contrast to the voyeurism of many walking-tour pedagogies, we experienced urban spaces of injustice and resistance in Bogotá in ways that encouraged each of us to consider and position our privileges. To understand the fear, love, empathy, and indignation that arose within us as we navigated the stark inequalities of the city as a learning collective.

We did occasionally meet in the Uniandes classroom, to watch a film or video clip or to review theoretical concepts. In this space, La Madre represented the very essence of community-driven knowledge production. Her occupation of the classroom, as a trans sex worker and activist in *el barrio* Santa Fe, contested every elitist foundation of the university, and disrupted the privileged sanctuary of many wealthier students, whose splendid isolation too often leads them to contribute to the reproduction of the classist criminalisation of poverty in Colombia. For example, during my time as a professor at Uniandes I witnessed multiple classist and sexist hate groups, including a Facebook group entitled "How to Identify a Poor Student at Uniandes" (*Como Identificar a un Pobre en Uniandes*).

One of my favourite lectures in the first iteration of our course introduced the concept of the "Right to the City." It was led by La Madre and her sister-in-struggle, Daniela Maldonado Salamanca (see opening quote), who she had invited to co-deliver the class. Together they presented their conception of what the right to the city may actually look like for trans women and sex workers. I remember that this conceptualisation held important critiques of the classist, segregated, and transphobic state that banishes trans bodies and sex work. The police state participates directly in erasure of trans lives or looks the other way when trans women are assassinated in its streets every day. During this lecture, La Madre and her sister-in-struggle opened a window on the world of the divided city in which the students live. I remember observing students' reactions as moments of radical consciousness building.

I remember admiring the love and deep friendship these women had developed while fighting on the frontlines for trans women's rights. They presented themselves as chosen family, as movement family, and as sisters for life. As these relationships transition and change over time, I reflect on the fragility of movement spaces and relationships and the dreadful consequences of capitalism.

Although I am unclear about the details, La Madre was accused by other community activists of resource mismanagement and NPIC-esque instrumentalisation of her sisters in the streets. La Madre distanced herself from her previous network of activism and has created a different initiative of trans street-level activism in a different zone of the city.

Movement Memories of La Tango: Fierce Maestra, Freedom Fighter in the Streets

La Tango has led justice-seeking campaigns in central Bogotá for many years, long before I arrived in 2008 and continuing today, including in Santa Fe, 18th street, and La Mariposa Plaza. She has contributed to the radical consciousness building of multiple generations of university-based, NGO-based, and street-level activists about the waves of social cleansing and state violence in the streets. She fought for the rights of the street-connected community of "La L" (the L-shaped community known by mainstream urbanites as "El Bronx"), a drug consumption zone in Bogotá. Her family was deeply connected to these spaces and La Tango herself grew up on the streets. La Tango grew up in one of the most violent and impoverished neighbourhoods of Bogotá as the daughter of street-connected parents. She accumulated heightened care and financial responsibilities in the household at a very young age. La Tango survived state violence, violence perpetrated in the sex work spaces, arbitrary arrest, and violence against the street-connected in city spaces such as "La L." La Tango also spent many years as a street vendor and advocate for this faction of the street-connected in Bogotá. As her street-level activism became more and more public facing, both academic and human rights sectors, her NPIC entanglement deepened. Despite her temporary accumulation of resources as her NPIC entanglement deepened, she has never forgotten the streets.

I met La Tango with an old friend and comrade, Timothy Ross, a British expat and cofounder of the Fenix Social Foundation, an organisation I volunteered with during my early years in Bogotá. I worked with La Tango during my predoctoral research and accompanied her throughout the city centre, visiting *ollas* (drug consumption zones in the streets) and multiple locations in the sex work zone in Santa Fe as I became a familiar face in the streets.

La Tango and I worked together from 2008 as part of different street "outreach" teams connected to multiple organisations we both volunteered with, including the Fenix Social Foundation and the Procrear Foundation. I accompanied La Tango in her daily activities, *recorriendo* [5] the city centre streets for hours and hours and then, with the beginnings of rainfall, seeking refuge in a cosy tango joint and dancing the night away.

La Tango eventually became one of the top sellers of Procrear Foundation's street-level magazine *La Calle*, a social entrepreneurship endeavour started by British journalist Henry Mance and later absorbed by the Procrear Foundation. La Tango also made a living selling pens in different parts of the city, most famously known in the university campuses of the city centre, including La Javeriana, Uniandes and El Externado, where she had also developed an important client base for *La Calle* magazine.

Years later La Tango and I continued our work together through the organisation PARCES, cofounded with students and other community members. Soon after La Tango launched a community-based organisation working with sex workers, drug users, and other street-connected individuals and received an Open Society Foundations grant, her father passed away. This left La Tango with impossible decisions regarding organisational resources as a person surviving in conditions of structural poverty, violence, and family tragedy.

La Tango recently reached out to me via WhatsApp, asking how I was doing amid the public health crisis of COVID-19. She said, "*sé que no merezco el amor de nadie, que soy una basura, que me merezco lo que me está pasando*"[6] (Personal communication, April 2020). To me, this statement illustrates the toxic consequences of the NPIC. A logic that slowly seeps into activists' veins, convincing them that they are nothing without it. That their fight is characterised by the size of their organisational budget or their visibility and acceptance in public spaces such as social media. By conforming with NPIC logics, well-meaning external funders unleashed a moral attack on this activist's desire to survive inside and outside the funding regime. Consequently, these funders undermined her mental health and sense of self-worth.

La Tango tried to take her life a few months ago, when everything in her organisation collapsed. This is now the second time I have experienced a sister-in-struggle's suicide attempt and epitomises for me the consequences of the NPIC. It sucks the life out of us. It tears apart our collectivity. It brings out the absolute very worst in everyone.

Discussion: Postcapitalism as Forgiveness of NPIC-Invoked Transgressions?

La Tango speaks about not deserving love, reducing herself to "*desecho*."[7] I argue that part of the imagined movement toward postcapitalist politics and dreams necessitates divesting from such simplistic reductions of human life and precarity. Activist-scholars proclaim ourselves to be advocates for the incarcerated, for decarceration, for the transformative potential of forward thinking and forgiveness. However, we also condemn those who act in individualistic lapses, amid capitalist chaos and extractive logics reproduced within the NPIC. In our own fight to support street-connected communities, do we not argue that no lives are expendable? Then how is it that our policing of each other's activism should lead to such destructive relations?

Based on both accounts shared above, I argue that we should not let positive movement memories slip away, or let productive contributions be erased as we are collectively dragged through the mud and eventually annihilated by the NPIC. These memories are part of our shared movement history. Everyone should have space to share these memories, from our unique positionalities and within the praxis of our individual and collective social memory. I dream for this space with my sisters. The structure of the academy keeps smothering the flame of this collective memory initiative through its neoliberal caging of our creativity and silencing of our affective analytics and voices. I want, more than anything, to keep dreaming with my sisters, of my sisters, to keep supporting my sisters as they shake off the shackles of the NPIC.

Does it not make some sense that individuals in marginalised communities will sometimes *take*, if they constantly have been *taken from*, if they have constantly been denied access to opportunities and even basic survival rights? What I argue for here reaches beyond a logic of redistribution. Of course, redistribution is part of it, in the sense that it seems reasonable for communities to reclaim what has always been theirs – for example, the surplus of billionaire investor and philanthropist George Soros' donations to Open Society Foundations (Open Society Foundations, n.d.). In a transformative justice frame, it makes good sense for marginalised communities to reclaim their knowledge, their dignity in academic and NGO spaces, their access to project funds, their access to the basics of human survival (such as food, shelter, and community space), and their access to dignified death and spaces of mourning as they bury their loved ones. As decolonial scholars such as Linda Tuhiwai Smith (1999) have argued for decades, this is a critical practice of "re-writing" and "re-righting" history. Here, individual acts of reclaiming and taking for survival purposes should not be seen as individual failure, but rather as symptoms of structural failure and inequality. Within the layers of complexity of this, I interpret the financial transgressions of my sisters-in-struggle as casualties of the NPIC. When ostracised by other activists in the community, the lives of La Madre and La Tango are both oversimplified, reduced to their identities as activists and decontextualised, replacing systemic failure with individual responsibility.

It is harrowing that social leaders are so quickly ostracised and shamed for mismanaging funds, by their movement family in the streets and by class-privileged activists in universities and non-profits. Activists who take a higher moral ground when observing their sisters' transgressions often miss the mark (Chatterton, 2006). These critics are quick to fuel viral character assassination campaigns and quickly lose sight of their sisters' humanity. They are quick to label mismanagement as corruption or theft, without situating these resources in the colonial and extractive origins of the global human rights funding giants who swoop in, give resources, and then pull out just as everything is burning down.

As a means of consciously avoiding the toxic tendency of social movement in-fighting and finger-pointing, we must understand that a large part of the gradual construction of two sisters like these as street-level enemies of the movement has to do with the handling of monies granted by powerful institutions external to the community (and in some cases external to Colombia). Many activists concluded long ago that the NPIC exists and that its funding is detrimental to the love and passion that keeps movements running forward (Incite!, 2017).

In their discussion of affects and emotions for a postcapitalist politics, J. K. Gibson-Graham (2006) argue that "[i]f we want to cultivate new habits of thinking for a post-capitalist politics, it seems there is work to be done to loosen the structure of feeling that cannot live with uncertainty or move beyond hopelessness" (p. 4). For me to reproduce the elite logics that unsympathetically eviscerate our sisters when they fail, is to treat the consequences of systemic failure as the failure of individuals, and to reproduce feelings of hopelessness.

By contrast, through this memory work that reclaims the positive personal histories of the movement and those who helped build it, I wish to respond to Gibson-Graham's provocation and propose that sisterhood, love, empathy, and forgiveness can be the beginning of what moves us beyond, or perhaps pulls us through, a post-capitalist reimagining of relationships within and between movements. What if movements think of friendship as something that moves us beyond the confines of this current world order into other worlds antithetical to constraint, limits, borders; worlds where we can imagine, dream, and move toward freedom? Movements held together by core principles of friendship, empathy, and love, which compel us to hold each other accountable in ways that are empathetic and generous. Movements that do not condone corruption or misuse of funds, but rather allow space to recognise that our failures are often a product of oppressive structures and inequality, which cannot be reduced to individual responsibility. Therefore, it is important not to succumb to the logic of the NPIC and the AIC or accept the current structure of justice-seeking work at face value. Rather, we in justice-seeking movements should push ourselves to dream forward and beyond these structures.

Conclusion

In part, imagining other worlds focuses on what individuals have done for society rather than the unsurprising extractive consequences of capitalism in street-level activists' work. What these activists have done to advance justice in the lives of their brothers, sisters, and comrades in the streets. When great transformations in society are remembered, why are street-level activists held to such different standards than other leaders? In dreaming forward, I also believe it is instructive to do the work of learning from our shortcomings as a movement family. Imagining other worlds and a post-capitalist afterlife of PAR also means imagining what our movement relationships would look like (or how we dream for them to be) in this afterworld.

In my attempt to imagine PAR in the afterlife of capitalism, I honour these activists and what I witnessed them contributing to the movement, before the NPIC and AIC co-opted and entangled their fierce, street-level fight for freedom.

Notes

1 In this chapter and elsewhere, I refer to sisters-in-struggle as individuals who have become like family in movement space in Colombia and beyond.
2 See *Sueños furiosos*, a trans political agenda publication: https://challengeinequality.luskin.ucla.edu/wp-content/uploads/sites/16/2019/09/Ritterbusch-Amy-SUE%C3%91OS-FURIOSOS.pdf. English translation of the quote by Olivia Miller, Master of Social Work student at University of California Los Angeles.
3 "Thrown into the abyss of the forgotten," a common Colombian saying.
4 *Peer leader* is a problematic term often used in non-profits to refer to "outreach" workers who visit their peers in the streets and invite them to become involved in the organisation's activities. The concept of "outreach" is also problematic in this sense, given that it reinforces the hierarchy and separation between human rights organisational life and street life.
5 Roaming through/working in the streets.
6 "I know I don't deserve anyone's love, that I am trash, that I deserve what is happening to me."
7 To trash, to waste, to the abject.

References

Askins, K. (2018). Feminist geographies and participatory action research: Co-producing narratives with people and place. *Gender, Place & Culture*, 25(9): 1277–1294. https://doi.org/10.1080/0966369X.2018.1503159.

Cahill, C. (2007). Repositioning ethical commitments: Participatory action research as a relational praxis of social change. *ACME: An International Journal for Critical Geographies*, 6(3): 360–373. https://acme-journal.org.

Cahill, C. (2010). 'Why do they hate us?' Reframing immigration through participatory action research. *Area*, 42(2): 152–161. https://www.jstor.org/stable/27801456.

Cahill, C., Sultana, F., & Pain, R. (2007). Participatory ethics: Politics, practices, institutions. *ACME: An International Journal for Critical Geographies*, 6(3): 304–318. https://acme-journal.org.

Chatterton, P. (2006). "Give up activism" and change the world in unknown ways: Or, learning to walk with others on uncommon ground. *Antipode* 38(2): 259–281. https://doi.org/10.1111/j.1467-8330.2006.00579.x.

Fals Borda, O. (1991). *Subversion and Social Change in Colombia*. New York: Columbia University Press.

Fals Borda, O. (2003). *Ante la crisis del país: Ideas-acción para el cambio*. Bogotá, Colombia: El Áncora Editores.

Fine, M., & Torre, M. E. (2006). Intimate details: Participatory action research in prison. *Action Research*, 4(3): 253–269. https://doi.org/10.1177%2F1476750306066801.

Gibson-Graham, J. K. (2006). *A Postcapitalist Politics*. Minneapolis: University of Minnesota Press.

Incite! Women of Color Against Violence. (2007). *The Revolution Will Not Be Funded: Beyond the Non-Profit Industrial Complex*. Durham, NC: Duke University Press.

Kelley, R. D. G. (2002). *Freedom Dreams*. Boston: Beacon Press.

Maldonado Salamanca, D. (2019). Manifiesto, 8 de Marzo, 2019 (Somos malas, podemos ser peores). In A. Ritterbusch & M. González (Eds.), *Sueños furiosos: Aportes para la construcción de una agenda política trans*. 9–11. https://challengeinequality.luskin.ucla.edu/wp-content/uploads/sites/16/2019/09/Ritterbusch-Amy-SUE%C3%91OS-FURIOSOS.pdf.

mrs c kinpaisby-hill (2011). Participatory praxis and social justice: Towards more fully social geographies. In V. M. Del Casino, M. Thomas, P. Cloke, & R. Panelli (Eds.), *A Companion to Social Geography*. 214–234. London: Wiley Blackwell.

Open Society Foundations (n.d.). Who we are. The Open Society Foundation. https://www.opensocietyfoundations.org/who-we-are.

Pigni, A. (2016). *The Idealist's Survival Kit: 75 Simple Ways to Avoid Burnout*. Berkeley: Parallax Press.

Ritterbusch, A. (2019). Empathy at knifepoint: The dangers of research and lite pedagogies for social justice movements. *Antipode*, 51(4): 1296–1317. https://doi.org/10.1111/anti.12530.

Ritterbusch, A., & González, M. (Eds.). (2019). *Sueños furiosos: Aportes para la construcción de una agenda política trans.* https://challengeinequality.luskin.ucla.edu/wp-content/uploads/sites/16/2019/09/Ritterbusch-Amy-SUE%C3%91OS-FURIOSOS.pdf.

Rodriguez. D. (2007). The political logic of the non-profit industrial complex. In Incite! Women of Color Against Violence (Eds.), *The Revolution Will Not Be Funded: Beyond the Non-Profit Industrial Complex*. 21–40. Durham, NC: Duke University Press.

Smith, L. T. (1999). *Decolonizing Methodologies: Research and Indigenous Peoples.* London: Zed Books.

Stoudt, B. G., Fox, M., & Fine, M. (2012). Contesting privilege with critical participatory action research. *Journal of Social Issues*, 68(1): 178–193. https://doi.org/10.1111/j.1540-4560.2011.01743.x.

Sultana, F. (2007). Reflexivity, positionality and participatory ethics: Negotiating fieldwork dilemmas in international research. *ACME: An International Journal for Critical Geographies*, 6(3): 374–385. https://www.acmejournal.org.

Torre, M. E. (2009). Participatory action research and critical race theory: Fueling spaces for nos-otras to research. *The Urban Review*, 41: 106–120. https://doi.org/10.1007/s11256-008-0097-7.

Torre, M. E., & Ayala, J. (2009). Envisioning participatory action research entremundos. *Feminism & Psychology*, 19(3): 387–393. https://doi.org/10.1177%2F0959353509105630.

Index

Note: Page locators in **bold** refer to tables and page locators in *italic* refer to figures.